Farming System and Sustainable Agriculture

NIPA GENX ELECTRONIC RESOURCES & SOLUTIONS P. LTD.
New Delhi-110 034

About the Author

Dr. Alok Kumar Patra graduated in Agriculture from Chiplima campus of Odisha University of Agriculture & Technology in 1988, obtained M.Sc. (Agriculture) in Agronomy from Institute of Agriculture, Visva Bharati, Santiniketan in 1990 and Ph. D. from Bidhan Chandra Krishi Viswavidyalaya, Mohanpur, West Bengal in 1996. He has an MBA degree from IGNOU, New Delhi.

Dr. Patra has published more than 90 research papers, 130 popular scientific articles and bulletins, eight books and 35 book chapters. He has 30 years of experience in research and teaching in UG and PG level on agroforestry and agronomy. Presently he is working as Chief Agronomist in the All India Coordinated Research Project on Integrated Farming Systems, OUAT, Bhubaneswar-751003.

Farming System and Sustainable Agriculture

Alok Kumar Patra
Professor of Agronomy
All India Coordinated Research Project on Integrated Farming Systems
Odisha University of Agriculture & Technology
Bhubaneswar-751003, Odisha

NIPA GENX ELECTRONIC RESOURCES & SOLUTIONS P. LTD.
New Delhi-110 034

NIPA GENX ELECTRONIC
RESOURCES & SOLUTIONS P. LTD.

101,103, Vikas Surya Plaza, CU Block
L.S.C. Market, Pitam Pura, New Delhi-110 034
Ph : +91 11 27341616, 27341717, 27341718
E-mail: newindiapublishingagency@gmail.com
www: www.nipabooks.com

For customer assistance, please contact
Phone: + 91-11-27 34 17 17
Fax: + 91-11-27 34 16 16
E-Mail: feedbacks@nipabooks.com

ISBN: 978-93-94490-62-8

Composed and Designed by NIPA.

Preface

The Green Revolution started in the early 1960s that led to the attainment of self-sufficiency in food grain production, is considered as the greatest agricultural transformation in the history of humankind. However, its benefits have been poorly distributed and thus, hunger still persists in some parts of the world. Moreover, the modern agriculture has raised several ecological issues, and put a question mark on the sustainability of various agricultural practices with high input use. Since resources on the earth are limited, there is a need to utilise these resources most judiciously and efficiently to harmonise ecology with prosperity. Now the prime focus of farmers, researchers and policymakers goes to sustainable development through the sustainable farming systems.

Keeping this in view, the Deans' Committee of Indian Council of Agriculture Research has recommended a course on Farming System and Sustainable Agriculture for undergraduate students of agriculture and horticulture. The present book is written as per the latest syllabus on the subject, which would be useful to the students.

Originality of the contents in this book is not claimed. In writing this book, a number of books and journals including international publications and university bulletins have been referred. I extend my sincere thanks to the authors and editors of these books and journals. Especially, I do express my gratitude to the authorities of Food and Agriculture Organisation of the United Nations for giving permission to use their literature freely. Every care has been taken to cite the bibliographic references. However, any omissions, misrepresentations, incorrect citations or other mistakes that may have occurred are regrctted. Any suggestion to improve the contents of the book will be highly appreciated. I take full responsibility for any errors in this book. Any shortcomings may be intimated so that it will be taken care of in the next edition.

I am grateful to my colleagues in the All India Coordinated Research Project on Integrated Farming Systems, Odisha University of Agriculture & Technology, Bhubaneswar for their help in various ways during the preparation of the manuscript. I thank to the New India Publishing Agency, New Delhi for bringing out the book timely and nicely.

I express my indebtedness and gratitude to my beloved parents who are a constant source of inspiration to me throughout my academic journey. I am also thankful to my wife Jharashree and daughter Prachurya for their constant support and encouragement.

April 04, 2022

Alok Kumar Patra
OUAT, Bhubaneswar

Contents

Glossary

3R approach: The principle of reducing waste, reusing and recycling resources and products is often called the '3Rs'.

Adaptation: The changes in policies, the changes in behaviour and the responses and solutions used to deal with the changing climate.

Additive intercropping: Here the plant population of the main or base crop remains unchanged as a pure crop stand but one or more minor crops are grown simultaneously with it by utilising the inter-row space.

Aggressivity: Aggressivity gives a simple measure of how much the relative yield increase in species 'a' is greater than that for species 'b' in an intercropping system.

Agricultural waste: The residues from production and processing of raw agricultural products such as fruits, vegetables, meat, poultry, dairy products and crops are called agricultural wastes.

Agrisilviaquaculture: A form of agroforestry consisting of tree (woody perennial), agricultural crop and freshwater aquatic animal components.

Agrisilviculture: A form of agroforestry consisting of tree and crop components.

Agrisilvipasture: A form of agroforestry consisting of tree, crop and pasture/ animal components.

Agrobiodiversity: The variety and variability of animals, plants and microorganisms that are used directly or indirectly for food and agriculture, including crops, livestock, forestry and fisheries. It comprises the diversity of genetic resources and species used for food, fodder, fibre, fuel and pharmaceuticals.

Agroecology: The study of bringing ecology principles into agriculture, including the diversity of species and genetics, recycling nutrients on the farm, and maintaining healthy soils.

Agroforestry: A collective name for land use systems and technologies where woody perennials (trees, shrubs, palms, bamboos, etc.) are deliberately used on the same land management units as agricultural crops and/or animals, in some form of spatial arrangement or temporal sequence.

Allelopathy: Any direct or indirect harmful effect that one plant has on another or mutually on each other through the production of chemical compounds that escape into the environment.

Alley cropping: A farming system in which arable crops are grown in alleys formed by trees or shrubs, established mainly to hasten soil fertility restoration and enhance soil productivity.

Annidation: Complementary interactions which occur both in space and time are referred to annidation.

Apisilviculture: In this agroforestry system, various honey or nectar producing trees frequently visited by honeybees are planted on the boundary of the agricultural field. Bees benefit the trees and trees in turn provide series of benefits to bees.

Aquaforestry: The aquaforestry system comprises of composite fish culture in farm ponds, and various trees and shrubs, leaves of which are preferred by fish are planted on the boundary and around fishponds. Leaves of these trees are used as feed for fish.

Aquaponics: A method of producing food that combines raising aquatic animals, such as fish or shrimp and plants. The waste from the animals is used as fertiliser for the plants.

Augmenting crops: When sub crops are sown to supplement the yield of the main crop, the sub crops are termed as augmenting crops.

Bench terrace: Conversion of a steep slope into a series of steps with near-horizontal hedges and near-vertical walls between hedges, using retaining walls (rock) or steep banks (soil) for intensive cropping.

Biochemical pesticide: These are naturally occurring substances that control pests by non-toxic mechanisms. Biochemical pesticides include substances that interfere with growth or mating, such as growth regulators, or substances that repel or attract pests, such as pheromones.

Biodiversity: It is the sum total of all the plants, animals, fungi and microorganisms in the world or in a particular area; all of their individual variation and all the interactions between them.

Bioherbicides: The uses of plant pathogens which are expected to kill the targeted weeds are called bioherbicides.

Biological farming: It is a system of crop production in which the producer tries to minimise the use of chemicals for control of crop pests.

Biological pest control: Biological control is a method of controlling insect pests and diseases using other beneficial organisms which rely on predation, parasitism, or some other natural mechanisms.

Biopesticides: Biopesticides are certain types of pesticides derived from animals, bacteria and trees. These have a narrow target range and a very specific mode of action. Biopesticides have no risk on environment.

Blind cultivation: In blind cultivation, the entire field is tilled shallowly, paying little attention to where the rows are. It is the easiest way to destroy the weeds that would be growing within the rows and presenting direct competition to the crop.

Carbon footprint: The quantity of greenhouse gas emissions created by a person, family, business or other entity.

Carbon sequestration (or terrestrial carbon sequestration or bio-sequestration): A process where airborne carbon dioxide is removed from the air and stored in plants as leaves, roots, stems, trunks and soils.

Carbon sink: A long-term storage reservoir for carbon, such as soil, wetlands, forests.

Carbon-stock: Carbon stored in vegetation or soil.

Carrying capacity: It is the theoretical equilibrium population size at which a particular population in a particular environment will stabilise when its supply of resources remains constant.

Catchment area: An independent unit of treated or untreated land area contributing runoff water to a reservoir.

Climate change: Climate change includes both human-induced global warming and its large-scale impacts on weather patterns.

Climate smart agriculture: This is a relatively new concept which was launched in 2009 advocating for better integration of adaptation and mitigation actions in agriculture to capture synergies between them and to support sustainable agricultural development for food security under climate change.

Compacted soil: Soils where the air pockets between soil particles have been reduced to such an extent that water can no longer infiltrate the soil, air is not held in the soil and plant roots are unable to be established in the soil.

Companion cropping: In companion cropping the yield of one crop is not affected by other. In other words, the yield of both the crops is equal to their pure crops.

Competition index: Competition of intercropped species for light, nutrients, water, carbon dioxide and other growth factors.

Competitive interaction: In this system the components interact in such a way that the increase in the yield of one component leads to decrease in the yield of other component due to competitive interaction.

Complementary interaction: In a system if the component crops help each other, by creating favourable conditions for their growth in such a way that the system provides a greater yield than the yield of their corresponding sole crops, then the interaction between the components is said to be complementary in nature.

Component crops: Individual crop species which are a part of the multiple cropping systems.

Compost: Organic material, such as leaves, stalks and roots, that has decomposed and is being added to soil as a fertiliser and to rejuvenate soil.

Conservation agriculture: This is an approach to managing agroecosystems for improved and sustained productivity, increased profits and food security while preserving and enhancing the resource base and the environment.

Conservation tillage: Methods for growing annual crops in the previous year's crop residue which reduce soil erosion and retain water and nutrients on the land; the farmers practising conservation tillage leave 30% of the crop residue on the fields.

Continuous grazing: Grazing animals on an entire pasture, rather than rotating the animals through subsections of the pasture

Contour: An imaginary line connecting points of equal elevation on the surface of the soil.

Contour farming: Contour farming is farming with row patterns that run nearly level around the hill.

Contour ridge: A series of parallel ridges on the contour of cultivated land which has grass or shrubs planted on them to control soil erosion and improve water management.

Cover crop: Crops grown primarily to cover the soil and to reduce the loss of moisture due to leaching and erosion by wind and water.

Crop diversification: Crop diversification is considered as an attempt to increase the diversity of crops through crop rotation, multiple cropping or intercropping with the aim to improve the productivity, stability and delivery of ecosystem services.

Crop equivalent yield: The yields from all the crops grown mixed or intercropped or sequentially cropped are converted to the yield of main crop of the system based on price of the produce.

Crop insurance: Crop insurance is a comprehensive yield-based policy meant to compensate farmers' losses arising due to production problems.

Crop rotation: Crop rotation refers to growing of different crops alternatively on the same piece of land in a definite sequence or process of growing different crops in succession on a piece of land in a specific period of time with an objective to get maximum profit from least investment without impairing the soil fertility.

Cropping intensity: Cropping intensity refers to raising of a number of crops from the same field during one agricultural year. It is the ratio of effective crop area harvested to the physical area expressed in percentage.

Cropping pattern: The yearly sequence and spatial arrangement of crops or of crops and fallow on a given area.

Cropping system: The cropping patterns taken up for a given piece of land, or order in which the crops are cultivated on a piece of land over a fixed period of time and their interaction with farm resources, other farm enterprises and available technology which determine their make-up.

Cultivated land utilisation index: This index is calculated by summing the products of land area planted to each crop, multiplied by the actual duration of that crop and divided by the total cultivated land area available during 365 days.

Cultural pest control: It includes crop production practices that make crop environment less susceptible to pests. Crop rotation, fallowing, manipulation of planting and harvesting dates, manipulation of plant and row spacing, and destruction of old crop debris are a few examples of cultural methods that are used to manage the pests.

Cuniculture: The agricultural practice of breeding and raising domestic rabbits as livestock for their meat, fur, or wool is termed as cuniculture.

Deficit irrigation practices: In arid and semiarid regions the irrigation strategies adopted for more effective and rational use of water are called deficit irrigation practices. Such management practices include regulated deficit irrigation, partial root drying and subsurface irrigation.

Deforestation: Deforestation is the purposeful clearing of forested land. Forests have been razed to make space for agriculture and animal grazing, and to obtain wood for fuel, manufacturing, and construction. Deforestation has greatly altered landscapes around the world.

Degraded pasture: Degraded pasturelands are native or planted pastures which have experienced a sharp decrease in carrying capacity, productivity and biomass production.

Double cropping: Cultivation of two crops in succession on a piece of land in a year.

Ecosystem: All the plants and animals in a given area and their physical environment, including the interactions among them.

Environmental pollution: Environmental pollution is the contamination of the physical and biological components of the earth/atmosphere system to such an extent that normal environmental processes are adversely affected.

Environmental sustainability: Environmental sustainability is the responsibility to conserve natural resources and protect global ecosystems to support health and wellbeing, now and in the future.

Erosion: The wearing away of the land surface by running water, wind, ice or other geological agents, including such processes as gravitational creep.

Farm: A farm is an economic unit in which crop and animal production is carried out with purpose of producing economic net returns.

Farmers market: A retail market where farmers sell their produce, meat and eggs directly to the consumers.

Farming: It is a process of harnessing solar energy in the form of economic plant and animal products.

Farming system: An appropriate combination of farm enterprises, viz. crops, livestock, forestry, fishery, apiary, poultry, and the means available to the farmer to raise them for profitability. It interacts adequately with the environment without dislocating the ecological and socioeconomic balance.

Feed grains: Grain crops that are used to feed livestock, such as corn, barley, and wheat.

Filler cropping: Growing of short duration crops in between the newly established perennial crops for few years to fill the space and to utilise the resources.

Fodder: Parts of plants which are eaten by domestic animals, these may include leaves, stems, fruit, pods, flowers or pollen.

Foliage: The mass of leaves of plants, usually of trees or bushes.

Food insecurity: A condition where a person does not have enough food to eat on a regular basis.

Food loss: Food is spoiled or discarded during production, processing, storage, and transportation phases.

Forage: Vegetative material in a fresh, dried, or ensiled state which is fed to livestock (hay, pasture, silage).

Global warming: Global warming is the long-term heating of earth's climate system observed since the pre-industrial period due to human activities, primarily fossil fuel burning, which increases heat-trapping greenhouse gas levels in earth's atmosphere.

Green manure: Green plant material incorporated into the soil to improve its fertility.

Greenhouse gases: A group of gases that hover in the atmosphere and that trap heat near the earth's surface, among them are carbon dioxide, methane, nitrous oxide, hydrofluorcarbons, perfluorcarbons and sulfur hexafluoride.

Groundcover: Living or non-living material which covers the soil surface.

Guard crops: The main crops are grown in the centre, surrounded by hardy or thorny crops such as safflower around pea or wheat with a view to provide protection to the main crop.

Hedgerow: A closely planted line of shrubs or trees, often forming a boundary or fence.

HEIA: In the high external input agriculture (HEIA), external inputs like chemical fertilisers, high yielding and hybrid seeds, pesticides, irrigation, mechanisation based on fossil fuels are extensively used.

Homegarden: A land use form on private lands surrounding individual houses with a definite fence, in which several tree species are cultivated together with annual and perennial crops, often with inclusion of livestock.

Indicators of agricultural sustainability: These are a composite set of attributes or measures that embody a particular aspect of agriculture.

Infiltration: The downward movement of water into the soil.

Integrated crop-livestock systems: Farms that include livestock on the landscape as well as crops; they are more beneficial in returning carbon to the soil and maintaining healthy soils.

Integrated pest management: This is an ecologically based approach to pest control that utilises a multidisciplinary knowledge of crop/pest relationships, establishment of acceptable economic thresholds for pest populations and constant field monitoring for potential problems.

Intensive agriculture: Intensive agriculture, also known as intensive farming, conventional, or industrial agriculture, is a type of agriculture, both of crop plants and of animals, with higher levels of input and output per unit of agricultural land area.

Intercropping: This refers to growing of two or more dissimilar crops simultaneously on the same piece of land.

Irrigation scheduling: Irrigation scheduling is the process used by irrigation system managers to determine the correct frequency and duration of watering.

Land degradation: Land degradation is a process in which the value of the biophysical environment is affected by a combination of human-induced processes acting upon the land. It is viewed as any change or disturbance to the land perceived to be deleterious or undesirable.

Land equivalent ratio: The relative land area under sole crop required to produce the same yield as obtained under a mixed or an intercropping system at the same management level.

Land utilisation index: Land utilisation index has been defined as the number of days during which the crops occupied the land during a year divided by 365.

Land-use system: The way in which land is used by a particular group of people within a specified area.

Leaching (of nutrients): The dissolving and washing away of nutrients down the soil profile by the action of rain water.

LEISA: The low external input sustainable agriculture (LEISA) is a way of farming, where it seeks to optimise the use of locally available resources by maximising the complementary and synergistic effects of different components of the farming systems.

Litter: Layer of decomposing plant material (leaves, branches, etc.) covering the ground, especially under trees.

Live mulching: Living mulch is a cover crop interplanted or undersown with a main crop, and intended to serve the purposes of mulching, such as weed suppression and regulation of soil temperature. Living mulches grow for a long time with the main crops, whereas cover crops are incorporated into the soil or killed with herbicides.

Microbial pesticides: Microbial pesticides are naturally occurring or genetically altered bacteria, fungi, algae, protozoans or viruses. These can be effective and used as alternatives to chemical insecticides.

Microclimate: The specific local climatic conditions near the ground or area around plants up to 2 m height, resulting from the modifications of the general climatic conditions by local differences in relief, exposure and cover, etc.

Microclimate amelioration: Microclimate amelioration is the difference between the local climatic conditions organisms are experiencing and their macroclimate, or the 'free-air' conditions of well-mixed air in nearby open areas.

Minimum tillage: Minimum tillage is a soil conservation system like strip-till with the goal of minimum soil manipulation necessary for a successful crop production. It is a tillage method that does not turn the soil over, in contrast to intensive tillage, which changes the soil structure using ploughs.

Mitigation: Mitigation involves strategies, processes and technologies that reduce greenhouse gases.

Mixed cropping: Growing of two or more crops simultaneously on the same piece of land, without any definite row arrangement is termed as mixed cropping.

Mixed farming: Cropping systems which involve the raising of crops, animals &/ or trees.

Mixed intercropping: Growing two or more crops simultaneously with no distinct row arrangement.

Mixed row intercropping: It is growing two or more crops simultaneously in the same piece of land intermingled within a distinct row arrangement.

Monoculture: The repetitive growing of the same sole crop on the same land.

Moriculture: Cultivation of mulberry plants is called moriculture.

Mulch: A layer of loose material on the soil to reduce moisture loss, moderate soil temperature and inhibit weed growth.

Multiple cropping: Growing two or more crops on the same piece of land in one calendar year is known as multiple cropping.

Multiple cropping index: This measures the sum of area planted to different crops and harvested in single year divided by the total cultivated area times 100.

Multipurpose trees: Trees yielding one or more products and offering environmental benefits as well.

Multistoried cropping (or multitier cropping): Growing plants of different height in the same field at the same time is termed as multistoried cropping. It is mostly practised in orchards and plantation crops for maximum use of solar energy even under high planting density.

Mushroom compost: Mushroom compost is a type of slow-release, organic plant fertiliser. It is the residual compost waste generated or left out after mushroom productions. It contains lots of salt and organic material along with enzymes and other nutrients that make it suitable habitats for different microbes, synergistically they are found to be beneficial in disease suppression and plant growth promotion.

Nitrogen cycle: The sequence of chemical and biological changes undergone by nitrogen as it moves from the atmosphere into water, soil and living organisms, and upon death of these organisms (plants and animals) is recycled through a part or all of the entire process.

Nitrogen fixation: The biological conversion of elemental nitrogen (N_2) to organic combinations or to forms readily utilised in biological processes, by nitrogen-fixing microorganisms. When brought about by bacteria in the root nodules of leguminous plants, it is referred to as symbiotic; if by free living microorganisms acting independently, it is referred to as non-symbiotic or free fixation.

No-till agriculture (or zero tillage): Keeping crop residues on the land and leaving the earth largely undisturbed during planting.

Nutrient pump: A deep tree root system, that takes up nutrients from deep soil layers and brings them to the surface in the tree and its litter fall.

Nutrient recycling: Nutrient recycling is a cyclic process that encompasses the movement of nutrients from the physical environment to living organisms and back to the environment.

Organic farming: Farming that does not use pesticides and artificial fertilisers but instead uses organic fertilisers and natural pesticide control. Also avoids using antibiotics and hormones for weight gain in animals.

Organic manures: These are organic materials derived from animal, human and plant residues which contain nutrients in complex organic forms.

Parallel intercropping: In this practice two crops are selected which have different growth habits and have a zero competition between each other and both of them express their full yield potential.

Partial root drying: Partial root drying is a new irrigation technique, where one half of the root system is allowed to drying conditions by withholding irrigation and the other half is irrigated. Wetted and dried sides of the root system alternate on a 7-14 day cycle. Partial root drying uses biochemical responses of plants to water stress to achieve balance between vegetative and reproductive growth.

Percolation tanks: Percolation tank is an artificially created surface waterbody, submerging in a highly permeable land, so that surface runoff is made to percolate and recharge the ground water storage.

Permaculture: A method of farming that involves caring for the earth and the earth's natural systems, providing for people to use the resources they need and returning waste to the earth's natural systems.

Plant genetic resources: Plant genetic resources are plant genetic materials of actual or potential value. They describe the variability within plants that comes from human and natural selection over millennia.

Pollinator: A bee, moth, butterfly or other insect, bird, bat, or other animal that moves pollen from the male anthers to the female stigma of flowering plants. The wind can also serve as a pollinator.

Precision agriculture: It involves applying pesticides and fertilisers to specific areas within a field based on the particular needs of the soils and plants of that area. It is based on sophisticated tools involving satellites and computer programmes and the concept that a farm field varies in things like soil type, elevation and water retention.

Pressure-state-response framework: The PSR indicators propose to evaluate the pressures of human activities on environmental states and to provide political and societal responses in order to come back to a desirable state.

Protective plant: Plants grown to protect crops, soil or land from adverse environmental factors.

Protein bank: Protein banks are blocks of forage plants deliberately planted to alleviate fodder shortages in arid, semiarid and mountainous regions, especially during the dry seasons.

Pyrolysis system: Pyrolysis is a process of chemically decomposing organic materials at elevated temperatures in the absence of oxygen. In pyrolysis systems, agricultural waste is heated up to a temperature of 400-600° C in the absence of oxygen to vapourise a portion of the material, leaving a char behind.

Rainwater harvesting: Rainwater harvesting is the collection and storage of rain, rather than allowing it to run off. Rainwater is collected from a roof-like surface and redirected to a tank, cistern, deep pit, aquifer, or a reservoir with percolation, so that it seeps down and restores the groundwater.

Ratoon cropping (or ratooning): It refers to raising a crop with regrowth coming out of roots or stalks after the harvest of the crop.

Real time contingency planning: Real time contingency planning is considered as any contingency measure, either technology related (land, soil, water, crop) or institutional and policy based, which is implemented based on real time weather pattern in any crop growing season.

Regenerative agriculture: Using farming techniques that enhance the land, including regenerating topsoil and increasing biodiversity; are resilient to climate change; that provide a livelihood for the farm families and the local community.

Regulated deficit irrigation: Regulated deficit irrigation is an optimising strategy under which crops are allowed to sustain some degree of water deficit and yield reduction.

Relative crowding coefficient: Relative crowding coefficient is a measure of the relative dominance of one component crop over the other in an intercropping or mixed cropping system. The coefficient (K) is determined separately for each component crop.

Relative economic efficiency: It refers to the capacity of the diversified cropping system for realising net profit in relation to the existing system and expressed in percentage.

Relative productivity efficiency: It refers to the capacity of the diversified system for production in relation to the existing system and expressed in percentage.

Relay intercropping: Growing two or more crops simultaneously during at least a part of the life cycle of each. A second crop is planted after the first crop has reached its reproductive stage but before it is ready for harvest.

Resilient agricultural practices: It is an approach that includes sustainbly using existing natural resources through crop and livestock production systems to achieve long-term higher productivity and farm incomes under any environmental change, especially climate variabilities.

Resource use efficiency: Resource use efficiency (fertiliser, water, etc.) is the output of any crop or anything else per unit of the resource applied under a specified set of soil and climatic conditions.

Risk assessment: Risk is an important aspect of the farming. Agricultural production risk derives from the uncertain natural growth processes of crops and livestock. Weather, disease, pests, and other factors affect both the quantity and quality of commodities produced.

Rooftop rainwater harvesting: This is the technique through which rainwater is captured from the roof catchments and stored in reservoirs. Harvested rainwater can be stored in subsurface ground water reservoir by adopting artificial recharge techniques to meet the household needs through storage in tanks.

Rotational grazing: A practice where grazing land is divided into segments called paddocks, with the grazing animals moved from paddock to paddock every few days. The forage grasses, both introduced and native grasses, are allowed to rejuvenate while the paddock is resting between grazing periods resulting in less soil erosion.

Rotational intensity: This is calculated by counting the number of crops in a rotation and multiplied by 100 and then divided by duration of the rotation.

Row intercropping: It is an intercropping where at least one crop is planted in regular rows, and crop or other crops may be grown simultaneously in row or randomly with the first crop.

Runoff: The portion of the precipitation on an area which is not absorbed by the soil but finds its way into the streams after meeting the persisting demands of evapotranspiration and other losses.

SDG (Sustainable development goal) 2.4.1: Ratio of area under productive and sustainable agriculture and agricultural land area.

Sequential cropping: Growing two or more crops in sequence on the same field. The succeeding crop is planted after the preceding crop has been harvested.

Sericulture: Sericulture or silk farming is the cultivation of silk through rearing of silkworm.

Shelterbelts: A wind barrier of living trees and shrubs established and maintained for protection of crop fields.

Shifting cultivation: It is a system of land use in which the site of cultivation is regularly changed, with older sites reverting to forest or bush fallow.

Silvipastoral system: A form of agroforestry systems consisting of the trees (woody perennial) and pasture/animal components.

Soil amendment: Materials added to the soil to make it more fertile such as compost, biochar, manure and artificial fertiliser.

Soil degradation: It refers to decline in the productive capacity of land due to decline in soil quality caused through processes induced mainly by human activities.

Soil organic matter: The organic fraction of the soil that includes plant and animal residues at various stages of decomposition, cells and tissues of soil organisms, and substances synthesised by the soil population.

Soil-water balance: Soil water balance is an account of all quantities of water added, removed or stored in a given volume of soil during a given period of time. The soil water balance equation thus helps in making estimates of parameters, which influence the amount of soil water.

Stover: The residue left on the fields after the grain is harvested, including stalks, leaves and roots.

Strip intercropping: It is growing two or more crops simultaneously in different strips on sloppy lands wide enough to permit independent cultivation, but narrow enough for the crop to interact with each other.

Substitutive intercropping: In the substitutive or replacement series of intercropping, the crops grown together are known as component crops or intercrops. Here, one component crop is introduced by the replacement of the other crop and no crop is sown with its fullest population as seeded in respective sole cropping. In this system, a definite proportion of a crop is sacrificed and the component crop is introduced in that place.

Supplementary interaction: If the two components interact in such a way that the yield of one component exceeds the yield corresponding to its sole crop without affecting the yield of other component, the interaction is known to be supplementary in nature.

Sustainability criteria: These refer to critical/threshold values against which the performance of each sub-indicator is assessed to classify the farm in terms of the sustainability level.

Sustainable agriculture: The management and conservation of the natural resource base and the orientation of technological change should be in such a manner as to ensure the attainment of continued satisfaction of human needs for present and future generations. Sustainable agriculture conserves land, water, and plant and animal genetic resources, and is environmentally non-degrading, technically appropriate, economically viable and socially acceptable.

Sustainable land use: Land use that maintains productivity of the land while conserving or enhancing the resources on which future production depends.

Sustainable soil management: Soil management is sustainable if the supporting, provisioning, regulating, and cultural services provided by soil are maintained or enhanced without significantly impairing either the soil functions that enable those services or biodiversity.

Sustainable value index: This index is used to assess the sustainability of a cropping system on monetary terms. In cropping systems where more than one crop is involved, the economic assessment is ideal one than the biological assessment. To work out the SVI, the monetary values of the economic produce are used instead of yield values. $SVI = \frac{Y - Sd}{Y\,max}$ where, Y is the estimated average net profit of a system over years, Sd is the estimated standard deviation and Ymax is the observed maximum profit among all the systems over years.

Sustainable yield index: This index is used to assess the sustainability of a cropping system or management practice which can be considered for wider adoption. It identifies the practices systems giving the maximum sustainable yield. $SVI = \frac{Y - Sd}{Y\max}$, where, Y is the estimated average yield of different practices or systems over years, Sd is the estimated standard deviation and Ymax is the observed maximum yield among all the systems over years.

Trap crops: These crops are grown in the main cropped field in definite rows to trap insect pests.

Triple cropping: Cultivation of three crops in succession on a piece of land in a year

Water use efficiency: It is a measure of the amount of biomass produced per unit of water used by a plant.

Windbreak: A strip of trees or shrubs or crop plants serving to reduce the force of wind and provide a protective shelter against wind.

1

Farming System and Their Types

A 'system' is a set of interrelated, interacting and interdependent elements operating together for a common purpose and capable of reacting as a whole to external forces. It is unaffected directly by its own output and it has a specified external boundary based on the inclusion of all significant feedbacks. A 'farm' is a system because several activities are closely related to each other by the common use of the farm labour, land and capital, by risk distribution and by the joint use of the farmer's management capacity. Thus, a farming system results from a complex interaction of interdependent and interrelated components of elements that bear upon the agricultural enterprises of the rural household. At the centre is the farmer who takes decision in an attempt to achieve his aspirations, goals and desired objectives within the limits of technologies and resources available to him. He uses inputs to get outputs in response to the technical elements which is the natural resource endowment in any given location restricting what the farming system can be. The human element provides the framework for development and utilisation of a particular farming system.

1.1 CONCEPTS OF FARMING SYSTEM

Farming system is a mixture of farm enterprises such as crop, horticulture, livestock, fishery, agroforestry and fruit crops to which farm family allocates its resources in order to efficiently manage the existing environment for the attainment of the family goal.

In any production system, some inputs are put to some processes or operations to get the desired outputs. Inputs are what go into the farm. There are two types of inputs, the natural and human inputs. In an agricultural system, the natural or physical inputs include weather and climate (precipitation, temperature, wind velocity, humidity, etc.), relief (height, shape and aspect), soil, geology and latitude. Farmers have little or no control over these. Changing the natural inputs can sometimes be done but it usually involves a lot of expenses. For example, areas with not enough rainfall get water from irrigation schemes, steep slopes can be cut into terraces and the climate can be greatly altered by using greenhouses. Examples of human inputs include seeds, fertilisers, pesticides, implements, machinery, livestock, animal feed, workers and buildings. These usually have to be paid for, although farmers can save some money by producing some of these themselves, e.g. grass is grown as a fodder crop and animals are bred within the system. Processes or operations are the actions that are carried out within the farm and allow the inputs to turn into outputs. These activities vary with the type of farm. Ploughing, sowing, weeding, adding manure and fertilisers, irrigation, harvesting and storage are important activities on an arable farm, whereas a major activity is milking on a dairy farm. Outputs are the things that are produced in the farm such as grains, vegetables, fruits, milk, meat, wool, eggs.

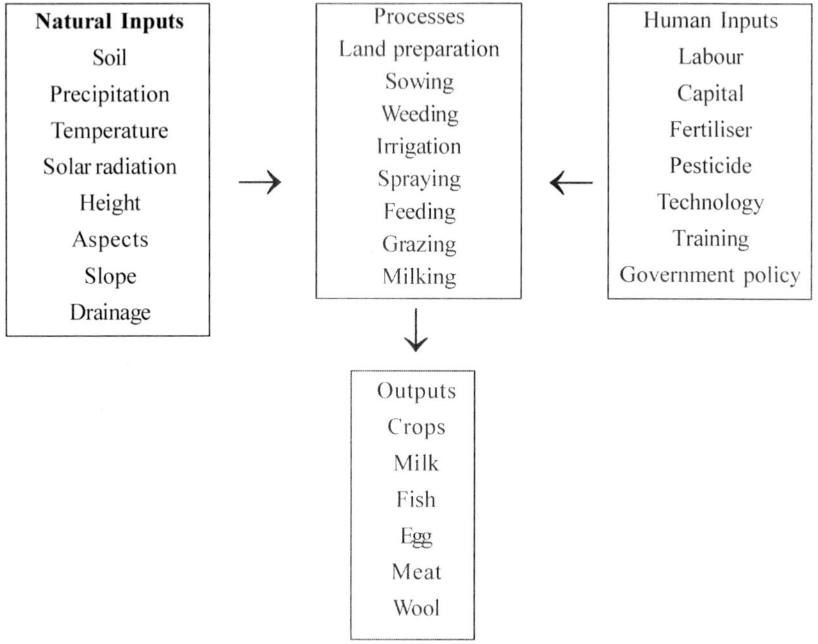

1.1.1 Definitions of Farming System

Different scientists have defined a farming system differently. However, many definitions, in general, convey the same meaning that it is the strategy to achieve profitable and sustained agricultural production to meet the diversified needs of farming community through efficient use of farm resources without degrading the natural resource base and environmental quality. Farming system is a decision making unit comprising the farm household, cropping and livestock system that transform land, capital and labour into useful products that can be consumed or sold (Fresco and Westphal, 1988).

Lal and Miller (1990) defined farming system as a resource management strategy to achieve economic and sustained agricultural production to meet diverse requirements of farm livelihood while preserving resource base and maintaining a high level of environment quality. On the other hand, a farming system is the result of complex interactions among a number of inter-dependent components, where an individual farmer allocates certain quantities and qualities of four factors of production, viz. land, labour, capital and equipments to which he has access (Mahapatra, 1994).

Farming system is a mix of farm enterprises such as crop, livestock, aquaculture, agroforestry and fruit crops to which farm family allocates its resources in order to efficiently manage the existing environment for the attainment of the family goal (Pandey et al., 1992).

Farming system is a set of agro-economic activities that are interrelated and interact with themselves in a particular agrarian setting. It is a mix of farm enterprises to which farm families allocate its resources in order to efficiently utilise the existing enterprises for increasing the productivity and profitability of the farm. These farm enterprises are crop, livestock, aquaculture, agroforestry and agri-horticulture (Sharma et al., 1991).

1.1.2 Principles of Farming System

Cyclic, rational and ecological sustainability are the three key principles of farming system.

- The farming system is essentially cyclic, e.g. organic resources-livestock-land-crops. Therefore, management decisions related to one component may affect the other.
- In any farming system, economic viability can be achieved through optimum allocation of available resources to various production systems and the use of these scarce resources in the most efficient way.
- Ecological sustainability can be obtained by integrating different components in such a manner that the by-products are utilised in the system with reduced negative environmental impacts.

1.1.3 Objectives of Farming System

Both production and ecological sustainability is the most important objective of a farming system. The following four specific objectives have been pointed out by Balasubramaniyan and Palaniappan (2010).

1. To identify existing farming systems in specific areas and assess their relative viability.
2. To formulate farming system models involving main and allied enterprises for different farming situations.

3. To ensure optimum utilisation and conservation of available resources and effective recycling of farm residues within a system.
4. To maintain sustainable production system without damaging resources and environment.

1.1.4 Scope of Farming System

Major farming enterprises are agronomic crops, vegetables, fruit crops, plantation crops, dairy, small ruminants, piggery, poultry, fishery, beekeeping, agroforestry, etc. Choice of the enterprises depends on many factors such as soil and climatic factors of the selected area, availability of resources such as land, labour and capital, present level of utilisation of resources, economics of the enterprises, technology available, managerial skill of the grower, etc. Practice of farming system has the following scopes.

1. Maximisation of yield of all components to provide stable income.
2. Biotic stress management through natural cropping systems practices.
3. Reducing the use of fertilisers, pesticides and other harmful agrochemicals to provide pollution free, healthy produce and environment to the society.
4. Increasing economic yield per unit area per unit time, profitability and sustainability.
5. Providing nutritious and healthy food for the farm family.
6. Creation of opportunity for effective recycling waste material.
7. Income or cash flow round the year.
8. Employment generation.
9. Increase in input use efficiency.
10. Improvement in standard of living of a farming community.
11. Always there is a scope for adoption of new technology.

1.1.5 Importance of Farming System

Environmental degradation, depletion in soil fertility and productivity, increased cost of farm inputs, fragmentation of holdings lead to unstable income and low profit from the farming. In this context, the farming system approach assumes great importance for sound management of farm resources to enhance farm productivity, reduce the degradation of environmental quality and improve the quality of life of farmers.

1. Farming system is considered a powerful tool for natural and human resource management. It enhances the farm input use efficiency.

2. It reduces the degradation of environmental quality and improves the quality of life of farming community.
3. Income through arable farming alone is insufficient for many of the marginal farmers. The other activities such as dairy, poultry, duckery, piggery, apiculture, aquaculture, etc. assume critical importance in supplementing their farm income.
4. Farming system maximises the productivity and net farm income on a sustainable basis. It is a process in which sustainability of production is the main objective.
5. This is a multidisciplinary whole-farm approach and very effective in solving the problems of small and marginal farmers.
6. The farming system approach creates employment opportunity throughout the year.
7. Livestock raising along with crop production is the traditional mixture of activities of the farmer. The nature and extent of farming may vary from region to region. It fits well with farm infrastructure, small landholdings and abundant labour force available with full utilisation of by-products.
8. Balanced food production is possible by integrating different enterprises in a farming system.
9. Farming system provides various models by integrating cropping with allied enterprises for irrigated, rainfed, hilly and coastal ecosystems.
10. Farming system achieves agro-ecological equilibrium through the reduction in the build-up of pests and diseases, through natural cropping system management and the reduction in the use of agro-chemicals.
11. Farming system saves energy through recycling of farm products and by-products. Also it provides fuel and timber for household requirement and infrastructure development, respectively, therefore reducing the pressure on natural forests.
12. Farming system meets fodder crisis during lean period.
13. Farming system supplies a number of raw materials to agroindustries.

1.2 CHARACTERISTICS OF FARMING SYSTEM

The term 'farming system', in its broadest sense, is any research that views the farm in holistic manner and considers interactions (between the components and of components with the environment) in the system (CGIAR, 1978). This type of research is most appropriately carried out by interdisciplinary teams of scientists who in association with extensionists, continually interact with the farmers in the identification of problems and in devising ways to solve them. It aims at generating and transferring technologies to increase the resource productivity for an identified group of farmers.

The farming system research activities are to be farmer-oriented, system-oriented with problem-solving and interdisciplinary approach. The farming system research also compliments mainstream disciplinary research, tests the technology in on-farm trials and provides feedback from the farmers. The strategy of farming system research should emphasise that the research agenda should be determined by explicitly farmers' needs through an understanding of the existing farming systems rather than its perception by the researchers. The farming systems research and extension should be dealt in holistic manner on farmers' participatory mode with problem-solving approach, keeping genders activity, interdisciplinary and interactive approach.

Farming system research is thus, an approach to agricultural research and development that view the whole farm as a system and focus on the interdependencies between the components under the control of members of the household and how these components interact with each other in respect of physical, biological and socioeconomic factors not under the household's control (Shaner et al., 1982). Many authors define characteristics of farming system research in different ways. The farming system core characteristics may be summarised as below (Behera, 2013; Kumar, 2013).

1. **It is problem-solving:** As an applied problem-solving approach, it emphasises on developing and transferring appropriate technologies to overcome the production constraints through the diagnosis of biophysical, socioeconomic and institutional constraints that influence the technological solutions.

2. **It is holistic:** The whole farm is viewed as a system encompassing interacting subsystems; and no potential enterprise is considered in isolation.

3. **It acknowledges the location specificity of technological solutions:** Recognising the location specific nature of agricultural production problems, it emphasises on testing and adaptation of technological solutions based on agroecological and socioeconomic specificities.

4. **It defines specific client groups:** Emphasis is made on the identification of specific and relatively homogeneous groups of farmers, with similar problems and circumstances for whom the technology is to be developed, as the specific client group. On the basis of common environmental parameters, production patterns and management practices, relatively homogeneous recommendation domains need to be identified.

5. **It is farmer participatory:** It revolves round the principle that successful agricultural research and development efforts should start and end with the farmers. Farmer participation is ensured at different stages of technology generation and transfer processes such as problem diagnosis, design and implementation of on-farm trials, and providing feedback through monitoring and evaluation.
6. **It gives weightage to ITK system:** The indigenous technical knowledge (ITK) which is time-tested at the farmers' level for sustainability through a dynamic process of integrating new innovations into the system as they arise, has to be properly understood by the scientists and utilised in their research activities.
7. **It is concerned with 'bottom-up' research strategy:** It begins with an understanding of the existing farming systems and the identification of key production constraints.
8. **It is interdisciplinary:** It lays greater emphasis on interdisciplinary cooperation among the scientists from different areas of specialisation to solve the agricultural problems that are of concern to farmer.
9. **It emphasises extensive on farm activities:** It involves problem analysis through diagnostic surveys, on-farm testing of the developed technologies, and providing feedback through evaluation to influence the research agenda of the research stations. It provides a structural framework for the farmers to express their preferences and apply their evaluation criteria for selecting technologies suiting to their circumstances.
10. **It is gender sensitive:** While explicitly acknowledging the gender-differentiated roles of farm family in agriculture, it emphasises the critical review of farming systems in terms of activities, analysis, access and control over the resources and benefits, and their implications in developing relevant research agenda.
11. **It is iterative:** Instead of trying to know everything about a system at a time, it requires step-by-step analysis of only key functional relationships.
12. **It is dynamic:** It involves recurrent analysis of the farming systems, permitting continuous learning and adaptations.
13. **It recognises interdependencies among multiple clients:** The generation, dissemination, adoption of relevant technologies to improve the productivity and sustainability of agriculture require productive and interactive linkages among the policy planners, scientists, developmental agencies and farmers. The approach attaches more importance for this critical factor.

14. **It focuses on actual adoption:** It is to be judged by the extent to which it influences the production of socially desirable technologies that diffuse quickly amongst specified groups of farmer clients.
15. **It focuses on sustainability:** It seeks to harness the strengths of the existing farming practices, and to ensure that the productivity gains are environmentally acceptable. Towards preserving the natural resource base and strengthening the agricultural production base, it attempts to develop technologies that are environment friendly and economically viable.
16. **It complements experiment station research:** It only complements, but does not substitute, the on-station research. It has to draw upon the scientific knowledge and technologies generated at research station. It has to be kept in mind that approach is not being promoted as panacea for all the maladies of local agricultural production systems.

1.3 TYPES OF FARMING SYSTEM

Farming systems may be classified on the basis of size of farm, land, labour and capital investment, income, water supply, type and intensity of rotation, degree of commercialisation, degree of grassland utilisation, cropping pattern, implements used for cultivation, etc.

1.3.1 Types of Farm Based on Size

According to size of the farm, farming systems can broadly be classified to collective or cultivation farming.

1.3.1.1 Collective farming

It is the direct collection of plant products from non-arable lands. Actual cultivation is not needed and the produce is directly harvested from uncultivated plants. Natural products like honey, gum, flower, silkworm cocoons are collected from forest area. Fish is also harvested from the natural waterbodies like river, lake and sea.

1.3.1.2 Cultivation farming

In this type of farming the farmers cultivate the land for growing crops and rearing livestock. There may be marginal, small scale or large scale type of farming.

Marginal farming

The marginal farmer does not always consider economic criterion in evaluating crop performance, because his first concern is food for his family. As such, he has nothing to market except on occasions when he has to make forced sale to get some cash for any specific need.

Characteristics of marginal farming

1. The farms or holdings are very small with greater pressure of population of the land.
2. The resource structure is very poor with the result that the farmer cannot give a proper direction to the allocation and utilisation of resources.
3. The products are consumed directly by the household and not sold in market.
4. The production factors are self-employed.
5. The price elasticity of production is small.
6. There is a complementary relationship between enterprises as some of them will have to be raised for by-products for cattle maintenance without consideration for loss or profit.
7. Price fluctuation of products has marginal effect on the production of such crops.

Small scale farming

In this case, farming is done on small size of holding and other factors of production are also small in quantity.

Advantages

1. Intensive cultivation is possible.
2. Labour problem does not affect the production.
3. It is easy to manage the farm.
4. Per unit area and per unit time output is more.
5. There is less loss due to biotic or abiotic stresses.

Disadvantages

1. Cost of production per unit area is more.
2. Mechanisation is not possible.
3. Farmer does not get employment round the year.
4. It is not possible to take advantages of various economic measures.

Large scale farming

Farming is done on large size holding with large amount of capital and labour force. More risk is also involved with large scale farming. In India, 40-50 ha land holding is said large scale farming whereas 100 ha farms are small farms in the USA, Australia and Canada.

Advantages

1. It is more economical since cost of production per unit is less.
2. Better marketing of agricultural products is possible.
3. Mechanisation on farm is possible. Costly machines like combined harvester, tractor, etc. can be maintained in the system.
4. Inclusion of subsidiary enterprises such as dairy, poultry, plantation, agroforestry, etc. are possible.
5. Post-harvest operations like processing, transportation, storage, packaging of produce are economical.
6. Production factors are most efficiently utilised.
7. There is often more than one component in the system. So if the farmer incurs loss from one enterprise under some adverse situation he gets some return from the other enterprise which reduces overall loss.

Disadvantages

1. There will be more loss in case the demand of produce decreases and/or the production exceeds the market demand which cause the fall of price.
2. A large farm will suffer huge loss in case of labour strike and natural calamities like flood, drought, insect pests or diseases.
3. It is difficult to manage the large scale farm than a small farm.

1.3.2 Types of Farm Based on Investment of Land, Labour and Capital

For any agricultural production system, land, labour and capital are the major inputs required. If a producer intends to increase his production level he has to bring more land under cultivation and accordingly more labour and more capital may be required.

1.3.2.1 Intensive cultivation

In intensive cultivation, more labour and more capital are invested on the land resource, without any increase in land area; in other words, land remains fixed in quantity while other factors are increased. Intensive cultivation is preferred to extensive cultivation under the conditions of population pressure and cheap labour availability. In earlier stage of development, more land was available due to low demographic pressure, and moreover, production technology was limited. Thus, extensive cultivation was adopted. However, with increase in human population and reduction in availability of land for cultivation, intensive cultivation becomes necessary and improvement in technology makes its adoption possible.

1.3.2.2 Extensive cultivation

When land is not a limiting factor and more area is brought under cultivation to increase the output it is termed as extensive cultivation. In extensive cultivation land is cheaply available but availability of other factors increases less proportionately. In earlier days when more land was available, extensive cultivation was in practice. In any country, both extensive and intensive cultivation practices go side by side for a certain time of period and afterwards, practice of intensive cultivation becomes important.

1.3.3 Types of Farm Based on Value of Products or Income

Based on income from the produce or the price offered to the produce, farming system can be of specialised farming, diversified farming, mixed farming, ranching or dryland farming.

1.3.3.1 Specialised farming

Under specialised farming 50% or more income is derived from one single source. In a specialised farming system, when the production of only one commodity is available for market, the farmer depends largely on a single source of income. A trend towards specialised farming is evident in areas where there are special market outputs and when economic conditions are fairly uniform for a long period. For example, a farm on which 50% or more of the receipts are from sugarcane would be called as sugarcane farm, and the one yielding 50% or more of its income from vegetables would be called a vegetable farm.

Advantages

1. The land is more efficiently used. It is more profitable to grow a crop on a land best suited to it. For example, jute is cultivated on a swampy land.
2. Better marketing for the produce is created. Specialisation allows better assembling, grading, processing, storing, transporting and financing of the produce.
3. Less equipment and labour are required. A fruit farmer needs only special machinery and comparatively less labour for raising fruits.

4. Farm mechanisation is possible in case of specialised farming system. The farmer can afford costly and efficient machinery. A combined paddy harvester can be maintained on a highly specialised paddy farm.
5. Efficiency and skill of the grower is increased. Specialisation allows a man to be more efficient and expert at doing a few things.
6. Better management of the farm is possible. The fewer enterprises on a farm are best managed and sources of wastage can easily be detected.

Disadvantages

1. There is a greater risk of failure of crop and market which together may ruin the farmer.
2. Productive resources like land, labour and capital are not fully utilised.
3. Fertility of soil cannot be maintained properly for lack of suitable rotation.
4. By-products of the farm cannot be fully utilised for lack of sufficient livestock on the farm.
5. There is no continuous cash flow. Farm returns in cash are generally received once or twice in a year.
6. Knowledge of various farm enterprises becomes limited.

1.3.3.2 Diversified farming

A farm on which no single product or source of income equals as much as 50% of the total receipt is called a diversified or general farm. On such a farm, the farmer depends on several sources of income.

Advantages

1. Better use of land, labour and capital is possible. The land is more efficiently used through adoption of crop rotations.
2. The family labour is utilised throughout the year.
3. More profitable use of equipment is obtained in diversified farming.
4. Business risk is reduced due to a crop failure or unfavourable market prices.
5. Regular and quicker returns are obtained from various enterprises throughout the year.
6. By-products of one enterprise can be used a production input in another enterprise.
7. Soil fertility is maintained due to crop rotation. Animal manures are also used to fertilise the crop components.

Disadvantages

1. Marketing is insufficient unless the producers arrange for the sale of their produce on cooperative basis.
2. Because of various jobs in diversified farming, a farmer cannot effectively supervise all the activities in a diversified farming system.
3. Better equipping of the farm is not possible because it is not economical to have expensive implements and machinery for each enterprise.
4. Some of the leaks in a farm enterprise may remain undetected due to diversity of operations.

1.3.3.3 Mixed farming

Mixed farming is a combination of crop production with rearing of livestock. It refers to that type of diversified agriculture in which a farmer invariably devotes to livestock production as a complementary enterprise. The most important reason for mixed farming is that it has been necessary in most of the regions to permit the use of a system of crop rotations combined with livestock enterprises, for getting draught animals for cultivation and also for maintaining and improving soil fertility. At least 10% of the gross income must be contributed by the livestock and the upper limit being 49% under Indian conditions. Bullocks are not usually considered as a part of the livestock enterprise, even then the farm can be called as mixed farming. However, a farm having cows, buffaloes, sheep, goat, poultry with crop production is called as diversified farming.

Advantages

1. It offers highest return on farm business as the by-products of farm are properly utilised.
2. It creates employment opportunity throughout the year. The family labour is used most efficiently.
3. Mixed farming promotes efficient use of land, labour and capital.
4. By-products of crop component are used as feed for livestock.
5. Manures are available from livestock to maintain soil fertility.
6. This farming system produces balanced food throughout the year.
7. Increases social status of farmer.
8. It provides greater chances of intensive cultivation.
9. It often gets higher returns on farm business.

Disadvantages

1. The management practices are often complicated. Primitive methods of growing crops and livestock are still in practice in a mixed farming in India.
2. Updated knowledge is required for all the enterprises. The farmer may not have that much skill and ability to manage a mixed farm.
3. The disposal of unproductive farm animals becomes a problem for the farmer due to religious sentiments and ethics.
4. Sometimes a good market is not available for the farm produce.
5. Labour requirement for the livestock component may be more.

1.3.3.4 Ranching

In this type of farming the livestock graze on natural vegetation. Ranch land is not utilised for tilling or raising crops. The word ranching does not come under the specification provided for the farm, i.e. it is not in the control of any owner nor is it enclosed by any boundary. Ranching means practice of grazing animals specially sheep and goat, and is always on public land. Sometimes, such land is utilised for raising dairy animals and it is known as dairy ranch. Ranching is very common in Australia and Tibet. In India, ranching is not common and is gradually disappearing because of the increasing pressure on agricultural land. Some parts of Rajasthan, ravines of river Chambal, etc. are the examples of this type of farming.

1.3.3.5 Dryland farming

Dryland farming generally refers to an area which receives less than 500 mm of annual rainfall. Areas where rainfall is up to 750 mm but is in coincidence with high temperature and greater wind velocity, resulting into a heavy loss of water may also be considered under this category. The crops are entirely depended on rainfall and soil moisture conservation practices are needed. Dryland farming needs attention on the following aspects for successful crop production.

1. Timely seedbed preparation for soil and water conservation and optimum crop stand establishment.
2. The crop fields should be weed free to prevent the transpiration of moisture through their leaves.
3. Organic manure should be adequately applied to improve water holding capacity of the soil.
4. Appropriate tillage practices are to be followed to increase the infiltration of rain as received by the fields.

5. The fields should be kept fallow for sometime so that they may save some of the water from one season for the growing of the crop in the next season.

1.3.4 Types of Farm Based on Water Supply

Based on water supply the farming may be rainfed or irrigated.

1.3.4.1 Rainfed farming

Growing of field crops which are entirely dependent on rain water for their water requirement is known as rainfed farming. Rainfed farming is risky with respect to water availability as the crop fully depends on monsoon. Success of a crop not only depends on total rainfall received but also its distribution during.

Characteristics

1. Kharif crop is sown with the onset of monsoon.
2. Short duration crops and varieties which withstand the moisture stress are more suitable for this farming.
3. Yield potential of crops is not fully exploited.
4. Desired crop rotation may not be followed. Only one or two crops can be taken per year.
5. Soils of these areas are deficient in plant nutrients. Applied fertilisers are not fully utilised by the crops.
6. Agronomic and engineering water conservation practices are adopted in the rainfed farming.
7. Mixed farming is preferred in rainfed areas.
8. Tree-based farming is encouraged for this region.
9. Use of organic mulch is a common practice.
10. Use of legumes as cover crops is encouraged.

1.3.4.2 Irrigated farming

In irrigated farming water is applied through external sources in addition to natural sources.

Characteristics

1. Crops are grown throughout the year. Intensive cropping is possible.
2. It is possible to grow cash crops like sugarcane.
3. It improves the yields of crops and gives good returns to the farmer.

4. It improves the groundwater storage as water lost due to seepage adds to groundwater storage.
5. Production inputs are utilised most efficiently.
6. Yield potential of crop varieties are fully exploited if other production inputs are not limited.
7. Desired crop rotation can be practised.

1.3.5 Types of Farm Based on Type of Rotation

There are different types of land use systems such as arable farming, tree farming, grassland use, etc. Rotation means the sequence of this basic type of land use on a given field.

1.3.5.1 Lay system

In this system, several years of arable farming are followed by several years of grassed and legumes utilised for livestock production.

Unregulated lay farming: In this system natural vegetation grasses, bushy pasture is allowed to grow during the period of fallow. This is an improved managed pasture.

Regulated lay system: During the period of fallow, certain types of grasses are grown or planted. These are the well-managed pasture with boundary fencing and adopting rotational grazing.

1.3.5.2 Perennial crop system

The crops which cover the land for many years (e.g. tea, coffee, rubber, oil palm, coconut, perennial fruit crops, sugarcane, etc.) altered with fallow with arable farming or grazing, etc.

1.3.6 Types of Farm Based on Intensity of the Rotation

Intensity of rotation is denoted by 'R' which is a simple and appropriate criterion for classification, and gives the true relationship between crop cultivation and total length of cycle.

$$R = \frac{\text{No. of crops grown in a field}}{\text{Years of rotation}} \times 100$$

R indicates the production of area under cultivation in relation to a total area available for arable farming.

1.3.6.1 Shifting cultivation

Because of shifting of field within an area the land is fallow for more years and a short period of cultivation. In this case R is less than 33%.

1.3.6.2 Fallow farming

Fallow farming is a technique in which arable land is left without sowing for one or more vegetative cycles. The goal of fallowing is to allow the land to recover and store organic matter while retaining moisture and disrupting the lifecycles of pathogens by temporarily removing their hosts. R is between 33 and 66%.

1.3.6.3 Permanent cultivation

When land is cultivated nearly every year is termed as permanent cultivation. Here large area is cultivated and small area is left fallow. R is >66%.

1.3.6.4 Multiple cropping

In this system more than one crop is cultivated on the same piece of land in a year. R is more than 100%.

1.3.7 Types of Farm Based on Degree of Commercialisation

Depending upon the quantity of farm produce sold in the market for earning money farming may be commercialised, partly commercialised or subsistence farming.

1.3.7.1 Commercialised farming

If more than 50% of the farm produce is available for sale, it is called commercialised farming.

1.3.7.2 Partly commercialised farming

If more than 50% of the value of produce is for home consumption, it is called partly commercialised farming.

1.3.7.3 Subsistence farming

Subsistence farming is a type of farming where the farmers cultivate the crop in their land for the living. Virtually there is a no sale of crop and animal products, but used for home consumption. The holdings are small in size. There is little scope for improved and mechanised cultivation practices. However, production resources and farm by-products are efficiently utilised. Family labour is used for crop cultivation and livestock rearing throughout the year.

1.3.8 Types of Farm Based on Degree Grassland Utilisation

This system involves the rearing of animals for economic production. It is classified on the basis of degree of nomadic.

1.3.8.1 Total nomadic

In this system the animal owners do not have permanent place of residence. They do not practise regular cultivation and their families move with the herds.

1.3.8.2 Semi-nomadic

Animal owners have a permanent place of residence near which supplementary cultivation is practised. However, for long period of time they travel with their herds to different grazing areas.

1.3.8.3 Transhumant

Under this system farmers with a permanent place of residence send their herds with herdsman for long period of time to different grazing areas.

1.3.8.4 Partial nomadic

Farmers have permanent place of residence and have herds at their disposal which remain in the vicinity.

1.3.8.5 Stationary animal husbandry

The animals remain on the holding or in the village throughout the entire year.

1.3.9 Types of Farm Based on Cropping Pattern

Farming system classified according to the leading crop and livestock activities of the holdings. When crop production is the major farm activity it is termed as crop-based farming system. When fishery is the major component in a farming system it is called pond-based farming system. Similarly, a farming system may be dairy-based, poultry-based, vegetable-based, fruit crop-based or tree-based farming system.

1.3.10 Types of Farm Based on Degree of Mechanisation

The mechanical operation on a farm is called farm mechanisation. It includes the use of manual implement, bullock drawn appliances and modern machines used in various farm operations like tractor ploughing, tube-well irrigation, harvesting and threshing by reapers and threshers, spraying by sprayers, etc. In post-harvest operations, mechanisation includes processing of products such as wheat or rice milling, cold storage, oil expelling, cane crushing and so on. The agroindustry corporations and some cooperative institutions provide custom service to farmers when needed. Based on the degree of mechanisation farming may be mechanised, semi-mechanised or non-mechanised.

1.4 FACTORS AFFECTING TYPES OF FARMING

The determinants of farming types can be grouped into the natural and the socioeconomic factors. The natural factors are comprised of the physical and the biological factors. These factors cause the type of farming to vary from area to area and result into a comparative advantage of producing a product in one region over the other. These factors do not change significantly from year to year.

1.4.1 Physical Factors

Physical factors include all external conditions and influences affecting the life and development of an organism.

Climate

It is the most important factor which affects the type of farming. It includes rainfall, length of growing season, temperature, solar radiation, relative humidity, frost, storm, wind velocity, etc.

Soil

The deep soils are generally well-drained and suitable for most of the crops. Light or shallow soils lose the moisture rapidly and so a large number of crops cannot be grown on such soils. The soil texture makes a soil heavy or shallow. If the soil particles are very small then the soil is clayey. Such soils are always slow to drain and become firm when dry. On the other hand, if the soil particles are very large then soil is sandy. Its water retention capacity is poor and it will not be suitable for most of the crops. Soil fertility is another important factor which determines the type of farming. In the case of fertile soils manurial cost will be less, thus reducing the cost of cultivation. Soil reaction, soil aeration, soil structure, presence or absence of growth-restricting substances is the other soil related factors affecting the type of farming.

Topography

Topography refers to the slope or height of the place where the farm is situated. Different crop may be grown at different elevations in hill areas. Thus changes in elevation affect the cropping pattern. At higher elevations, the temperature is low and growing seasons are generally shorter and thus are more suitable for orchards, tea plantations, potato cultivation, etc. As the levelled land in such regions is scarce and that too scattered, the farms are generally small in comparison to those of plains. The soil is rocky with various levels of fertility. Heavy machinery cannot be utilised at high altitude because of the rough and sloppy surface of the fields which happen to be small and scattered. This also affects the type of farming.

1.4.2 Biological Factors

Biological factors such as crops, livestock, weeds, pests and diseases also affect the type of farming of a region. Natural vegetation and adaptation ability of crops to a particular region decide the cropping pattern of that area. Composition of dairy animals (cows and buffaloes), small ruminants (goat and sheep) and poultry birds (chicken, duck, turkey, quail, guinea fowl, pigeons, etc.) in a farm decide the type of farming to be followed.

Prevalence of the pests such as weeds, diseases and insects in a farm also affect the type of farming. The occurrence of pests and diseases on an endemic fashion in a region will ruin or extinguish the entire type of farming and gives rise to new type of farming. In Guntur and Prakasam districts of Andhra Pradesh during 1980s the attack of white fly on cotton caused a 40% decrease in cotton farming area. The cultivation of pulses and oilseeds emerged as a new type of farming in this cotton belt.

1.4.3 Socioeconomic Factors

Economic and social factors determine the crop and livestock enterprises to be taken up in a particular region within choice established by physical factors.

Marketing costs

The cost of marketing farm products and the marketing problems determine what products will or will not be produced. The producer's share in consumer's price will decide the method of sale and choice of products by the producers. A low share to the producers will cause little preference to what products in relation to high producers share.

Changes in relative value of farm products

The response of the production of a particular product is much influenced by the farm product-price fluctuations and profitability, determining in turn the type of farming. Especially, the area and production of cash crops like cotton, tobacco, sugarcane, etc. are prone to change in a cyclical manner adjusting to the price fluctuation and profitability.

Availability of labour and capital

Farm enterprises requiring intensive operations require a good amount of labour involvement and capital investment. Labour and capital intensive farming in a region brings about migration of labour from one region to other region. Access to credit is a determining factor for farming types adopted by resource-poor farmers.

Land value

Low land values attract the enterprising farmers from land value areas, to settle and develop a new type of farming in those areas. Due to this new type of farming, low land values will go up in the course of time.

Competition between enterprises

Specialisation or diversification of farming is decided by the competition among the enterprises depending upon their relative profitability and resources use competition.

Cycles of over- and under-production

Agricultural enterprises are frequently subject to cycle of over- and under-production resulting in surplus/scarcity of production and low/high prices. This results in speculation and uncertainty in the type of crops to be grown. As a result, farmers prefer to grow such crops whose prices are on the rise or stable.

Choice of farming group

The type of farming or choice of products sometimes is determined by the personal preferences of the farmers, either due to traditional values or attachments they had in that particular crop. Family composition, food preferences, health issues and nutrition requirement of the people, education level of the farmers, gender relations, etc. also have effect on the type of farming in a region.

Risk aversion

Many farmers live on the edge of extreme uncertainty, sometimes falling just below, and sometimes rising just above the threshold of survival. Farmers do not know whether rainfall will be good or bad over a season; they do not know the prices they will receive for produce sold; and they do not know whether their crops will be infected by disease. These risks are not under the control of farmers. Attitudes to risk are often related to the financial ability of the farmer to accept a small gain or loss. Generally, the farmers (especially, marginal and small) are the most risk averse. Thus sometimes the type of farming is affected by the risk bearing ability of the farmers.

Off-farm opportunities

Off-farm work is any activity undertaken by the farmer or farm household outside farming as an additional source of income. Some farmers divert themselves from farming if off-farm income opportunities are available. Thus this affects the type of farming.

Technology, input supply and extension

Sophisticated technologies are now available to make the agriculture more profitable, efficient, safer, and more environmentally friendly. The agricultural extension service provides information inputs and knowledge on agricultural practices to farmers to increase agricultural productivity. Timely availability of production inputs is essential for success of any farming. Thus, these factors affect the types of farming.

1.5 SYSTEMS OF FARMING

The term 'system of farming' is generally referred to the methods of agriculture and the type of ownership of land. If the farming has been classified on the basis of economic and social functioning, it is called as systems of farming.

1.5.1 Cooperative Farming

A cooperative farming society is one in which members pool their land voluntarily and manage it jointly under a democratic constitution. It is an essential element of cooperative farming that its constituent members agree and surrender their individual rights and capacity to take major decisions in respect of farming enterprises to a common body constituted by them and accept its decision. There is a distinct difference between individual farming on cooperative lines and cooperative farming. In case of the former the individual farmer is in full possession of his land and himself carries on the basic operations of cultivation, he joins a cooperative body only for the sake of getting those services which are beyond his individual capacity while in the latter, the farmer transfers possession as well as ownership of his farm to a cooperative society.

The cooperative farming necessarily applies to the pooling of land and joint management. The working group on cooperative farming defines cooperative farming society as a voluntary association of cultivators for better utilisation of resources including manpower and pooled land and in which majority of the members participate in farm operation with a view to increasing agricultural production, employment and income. There are four types of cooperative farming societies in India. Out of these better farming and tenant farming societies have developed in the form of service cooperatives. But, it is distinguished from cooperative farming because there is no pooling of manpower and land in these societies and their members take this responsibility and risk of farm operations individually.

Cooperative better farming

These societies are based on individual ownership and individual operation. The farmers who have small holding or limited resources join to form a society for some specific purposes, e.g. use of heavy machinery, sale of products, etc. They are organised with a view to introduce improved methods of agriculture. Each farmer pays for the services which he receives from the society. Profit is not distributed, the earnings of the members from his piece of land, after deducting the expenses, became his profit.

Cooperative joint farming

It means the pooling of land and other possible resources. The members form a general body which formulates the schemes and does the duties of administration. The ownership is individual but the operations are collective. The management is democratic and is elected by the members of the society. A member receives daily wages for his daily work and the profit in the end is distributed according to his share in land.

Cooperative tenant farming

Such societies are usually organised by landless farmers. In this system land belongs to the society. The land is divided into plots which are leased out for cultivation to individual members. The tenants have no right on land but they carry on this business independently. The society arranges for agricultural requirements such as credit, seeds, manures, marketing of the produce, etc. A tenant gets all the income after deducting the rent of land and charges for other services provided by the society.

Cooperative collective farming

Both ownership and operations under this system are collective. Members do not have any right on land and they cannot take farming decisions independently but are guided by a general body which is supreme. It undertakes joint cultivation for which all members pool their resources. Profit is distributed according to the labour and capital invested by the members.

1.5.2 Peasant Farming

This system of farming refers to the type of organisation in which an individual cultivator is the owner, manager and organiser of the farm. He makes decisions and plans for his farm depending upon his resources which are generally meagre in comparison to other systems of farming. The biggest advantages of this system is that the farmer himself is the owner and therefore to take all sorts of decisions. A general weakness of this system is that the resources with the individual are less. Another difficulty is because of the law of inheritance, an individual holding goes on reducing as all the members in the family have equal rights in that land.

1.5.3 State Farming

State farming, as the name indicates, is managed by the government. Here land is owned by the state and the operation and management is done by the government officials. The state performs the function of risk bearing and decision making, where cultivation is carried on with help of hired labour. All the labourers are hired on daily or monthly basis and they have no right in deciding the farm policy. Supervision is done by the farm manager or farm in-charge. Such farms are not very paying because of lack of incentive. Farm policy is usually planned at the top whereas farming is such a profession which requires immediate and at the spot decision. There is no dearth of resources, at such farms but sometimes it so happens that they are not available in time and utilised fully.

1.5.4 Capitalistic Farming or Estate Farming

The management and ownership of such farms is under rich persons or capitalists, corporations or syndicates. The size of such farms is sufficiently large and the management is also quite efficient. Capital is supplied by one or a few persons or by many, in which case it runs like a joint stock company. In such farms, the work is carried on with hired labour. Resources are plenty, latest technical knowhow is used. Management is paid and the general policies are decided by managing body or board of directors. Sugar factories farms, rubber, coffee and tea plantations are some examples of such a system in India. These farms are very common in USA, Australia and Canada. The advantages of such farming are good supervision, strong organisational setup, sufficient resources, etc. Its weaknesses are that it creates socioeconomic imbalances and the actual cultivator is not the owner of the farm.

1.5.5 Collective Farming

The name, collective farming implies the collective management of land wherein large numbers of families or villagers residing in the same village pool their resources, e.g. land, livestock, machinery, etc. A general body having highest power is formed which manages the farm. The resources do not belong to any family or farmer but to the society. If any farmer wants to dissociate from it, he can do so, but he cannot go with his share. Money is lieu of his share will be given to him. This system had started in erstwhile USSR. The worst thing with this system is that the individual has no voice. Farming is done generally on large scale, therefore is mostly mechanised. This system is not prevalent in India but common in communist countries.

1.5.6 Contract Farming

One more farming system is emerging in India that is contract farming. In this system land is taken on long time lease base from individual owners. Land is remained under ownership of individuals but they have no role to play in farm policy. They are paid rental value of land. It is a good system in developed countries where holdings are large. But in India 75% farmers are small and marginal and if they leased out their land, they will be forced to migrate to cities in search of jobs. In long-term, there are chances of landlessness also.

Table 1.1. Ownership and operation in different farming systems.

System of farming	Type of ownership	Types of Operation
Cooperative better farming	Individual	Individual
Cooperative joint farming	Individual	Collective
Cooperative tenant farming	Collective	Individual
Cooperative collective farming	Collective	Collective
Collective farming	Society/state	Society/State
Capitalistic farming	Individual	Individual
State farming	State	Paid management
Peasant farming	Individual	Individual

1.6 FACTORS AFFECTING THE SYSTEM OF FARMING

The major factors which have an impact on the selection of various systems of farming mainly relate to holding size, volume of business, availability of resources, capability of using the resources properly and utilising the facilities given by the Government and other institutions.

Size of land holdings

If the size of holding is such that it provides enough income to meet the requirements of the farmer and his family and also provide enough work for them, the peasant farming may be preferred. But in the case of very small size of holdings, it may be better to pool the land and work together in the form of a cooperative society or collective farming. It will enable the farming community to undertake some off-farm activities to earn additional income. Very small holdings often result in an uneconomical use of resources as the resources which a farmer possess for farming operations may remain idle for most of the time. For example, bullock power may remain idle or surplus for most of the period during the year because of lack of work. Therefore, for proper and full utilisation of bullock power, the bullocks should remain fully engaged and that is possible only when the size of the farm is adequate. If small holdings are pooled together, then the pooled land will give the advantage of large farms.

Volume of business

Volume of business is a broad term than the size of holding. Besides, the size of holding, the volume of business also indicates the number of enterprises taken up in a year, the total output produced, requirement of manpower and net income obtained. For making the business a success, it should be run in a proper way as to get advantage of large-scale production. Therefore, collective or cooperative system of farming may prove better if the individual farms do not provide a desirable volume of business.

Availability and use of resources

Persons who have comparatively large amount of capital may adopt capitalistic system of farming making use of capital intensive practices of production. They may purchase costly implements which require more capital in the beginning but prove to be cheaper in terms of annual cost to be incurred on them due to the prompt and efficient service they provide.

Availing of facilities

Some facilities such as credit facility and marketing facilities which individual farmers require but cannot avail of due to the small size of business, may be availed of by adopting a specific system of farming like cooperative farming.

2

Farming System Components and Their Maintenance

In agriculture, crop production is the main activity. However, the income obtained from crops may hardly be sufficient to sustain the farm family throughout the year. Assured regular cash flow is possible when the crop is combined with other enterprises. These other enterprises or components are called allied components. There are several allied components in a farming system. Judicious combination of enterprises, keeping in view of the environmental conditions of a locality will pay greater dividends. The allied components include dairy, goatery, fishery, poultry, duckery, agroforestry, beekeeping, mushroom cultivation, sericulture, vermicompost, etc.

2.1 CROP PRODUCTION

The crop activities consist of field crops, vegetable crops, fruit crops, plantation crops and tree crops.

Cereals

Cereals are the cultivated grasses grown for their edible starchy grains. Generally grains of cereals such as rice, wheat and maize are used as staple food.

Millets

Millets are the small grained cereals which are of minor importance as food. However, now these are gaining importance at national level for their nutritional values. Millets are used as staple food in drier regions. Major cultivated millets in India include finger millet (ragi), foxtail millet, sorghum millet (jowar), pearl millet (bajra), little millet and barnyard millet.

Pulses

Seeds of leguminous crop plants are used as food. These are rich in protein. Major pulse crops are greengram, blackgram, chick pea, pigeon pea, field pea, and lentil.

Oilseeds

Crop seeds that are rich in fatty acids are used to extract vegetable oil for human consumption or industrial uses. Major oilseed crops include groundnut, sesame, rapeseed and mustard, sunflower, safflower, soybean, linseed, niger and castor.

Forage crops

This refers to crop vegetative biomass, fresh or preserved, used as feed for livestock. Forage crops include berseem, cowpea, rice bean, stylo, field bean, cluster bean, lucern, anjan grass, napier, guinea grass, etc.

Fibre crops

Fibre may be obtained from seed such as cotton; stem or bark such as jute, mesta, sunhemp and flax; leaf such as agave and pineapple.

Sugar and starch crops

Crops grown for the production of sugar and starch are sugarcane, sugar beet, potato, sweet potato, tapioca, etc.

Spices and condiments

Crop plants or their products are used to add flavour, taste and zest, and sometimes colour to the fresh or preserved food. The major crops under this category are ginger, garlic, fenugreek, cumin, turmeric, chilli, onion and coriander.

Medicinal crops

These crops are grown for use in preparation of medicines. Mint, *Aleo vera,* basil, calendula, rosemary and lavender are some of the medicinal crops.

Beverages

Products of these crops are used for mild, agreeable and stimulating liquors meant for drinking such as tea, coffee and cocoa.

Vegetable crops

Vegetable crops include cucumber, tomato, okra, pumpkin, eggplant, water melon, sweet potato, carrot, radish, lettuce, spinach, cabbage, cauliflower, broccoli, pumpkin, bottle gourd, ridge gourd, pointed gourd, potato, sweet potato, yam, etc.

Fruit crops

Short duration fruit crops like papaya, banana, citrus and pineapple are suitable for any farming system. Fruit crops represent a wide range of woody perennial species cultivated in orchards where soils vary greatly in their biological, chemical, and physical properties. Mango, guava, jackfruit, litchi, sapota, apple, ber, wood apple, jamun, etc. are the major fruit crops in India.

Plantation crops

Areca nut, cardamom, cashew, cocoa, coconut, coffee, oil palm, rubber, and tea are the major crops grown on a plantation scale. Most of these plantation crops are cultivated as monocultures, exceptions being coconut, areca nut, cocoa, and coffee, which are grown either as a mono crop or as a multiple crop.

Tree crops

Tree crops include teak, acacia, sissoo, poplar, bamboo, etc. These are grown on farm boundary or as block plantation. Tree-based farming systems are recommended for dryland areas. Integration of tree crops in farming system reduces the pressure on natural forests.

2.2 DAIRY FARMING

Dairying has become an important secondary source of income for millions of rural families and has assumed the most important role in providing employment and income generating opportunities particularly for marginal and women farmers. Most of the milk is produced by animals reared by small, marginal farmers and landless households. However, a chronic shortage of cattle feed coupled with the poor quality of fodder has become the major constraint in dairy farming in India.

2.2.1 FAO Guidelines on Dairy Farming

The objective for good dairy farming practice is that safe and quality milk should be produced from healthy animals using management practices that are sustainable from an animal welfare, social, economic and environmental perspective. Food and Agriculture Organisation of the United Nations and the International Dairy Federation have issued the following guidelines on animal health, milking hygiene, nutrition (feed and water), animal welfare, environment and socioeconomic management aspects to achieve the desired outcome from a sustainable dairy farming (FAO & IDF, 2011).

Animal health

1. Establishment of the herd with resistance to disease
2. Prevention of entry of disease onto the farm
3. Effective herd health management programme in place
4. Use of chemicals and veterinary medicines

Milking hygiene

1. Prevention of injuries to the animals during milking
2. Hygienic conditions for milking
3. Proper handling of milk after milking

Nutrition

1. Feed and water supplies from sustainable sources
2. Feed and water of suitable quantity and quality
3. Storage conditions of feed
4. Traceability of feedstuffs brought onto the farm

Animal welfare

1. Free from thirst, hunger and malnutrition
2. Free from discomfort
3. Free from pain, injury and disease
4. Free from fear

Environment

1. Implementation of an environmentally sustainable farming system
2. Appropriate waste management system
3. Impact of dairy farming practices on the local environment

Socioeconomic management

1. Effective and responsible management of human resources
2. Carrying out farm tasks safely and competently
3. Management of the enterprise to ensure its financial viability

2.2.2 Breeds of Dairy Animals

There are 50 well-defined breeds of cattle and 17 breeds of buffaloes in India.

2.2.2.1 Indigenous cattle breeds

Indigenous breeds are well adapted to the agroclimatic conditions of the country and are resistant to many tropical diseases and can survive and produce milk on poor feed and fodder resources. Some of these breeds are well known for their high milk and fat production. However, the production potential of these animals has deteriorated over a period of time due to lack of selection. The high producing exotic breeds do not have the above characteristics and are very difficult to manage in tropical Indian scenario. Thus, systematic efforts are required for the genetic improvement of these breeds for milk production.

Table 2.1. Important indigenous cattle breed for milk production

Breed	Breeding tract	Characteristics and yield potential
Sahiwal	Punjab, Rajasthan	Also known as Lola, Lambi Bar, Montgomery, Multani and Teli. Milk yield 2725-3175 kg per lactation period of 300 days.
Red Sindhi	Odisha, Tamil Nadu, Bihar, Kerala, Assam	Also known as Red Karachi and Sindhi. Milk yield 1100-2600 kg per lactation. Widely used in crossbreeding programmes. First calving at 39-50 months and intercalving period 425-540 days.
Gir	Gujarat	Also known as Gujarati, Kathiawari, Sorthi and Surati. Milk yield 1200-1800 kg per lactation. First calving at 45-54 months and intercalving period 515-600 days.
Rathi	Rajasthan	Average milk yield 1560 kg per lactation, ranging from 1062 to 2810 kg.
Shweta Kapila	Goa	Also known as Gaunthi and Gavthi dhavi. Average milk yield 510 kg per lactation ranging from 350 to 650 kg with an average milk fat 5.21%, ranging from 4.5 to 6.4%.
Badri	Uttarakhand	Also known as Pahadi. Average milk yield 632 kg per lactation, ranging from 547 to 657 kg with an average milk fat of 4%, ranging from 3.6 to 4.4%.
Belahi	Haryana	Also known as Morni. Average milk yield 1014 kg per lactation ranging from 182 to 2092 kg with average milk fat 5.25%, ranging from 2.37 to 7.89%
Binjharpuri	Odisha	Milk yield per lactation ranges from 915-1350 kg with milk fat ranging from 4.3 to 4.4%.
Deoni	Maharashtra, Karnataka	Also known as Surti, Dongarpati, Dongri, Wannera, Waghyd, Balankya and Shevera. Average milk yield per lactation 868 kg, ranging from 638 to 1229 kg with an average milk fat is 4.3%, ranging from 2.5 to 5.3%.

Gangatiri	Bihar, Uttar Pradesh	Also known as Eastern Hariana or Shahabadi. Average milk yield 1050 kg per lactation, varying from 900 to 1200 kg with an average fat of 4.9%, varying from 4.1 to 5.2%.
Gaolao	Maharashtra, Madhya Pradesh	Also known as Arvi and Gaulgani. Average milk yield 604 kg per lactation, ranging from 470 to 725 kg with 4.32% fat.
Hariana	Haryana	Average milk yield 997 kg in a lactation, ranging from 693 to 1745 kg. Age at first calving 40-60 months.
Himachali Pahari	Himachal Pradesh	Also known as Gauri and Himdhenu. Average milk yield 538 kg per lactation, ranging from 300 to 650 kg with average milk fat 4.68%, ranging from 4.06 to 5.83%.
Kankrej	Gujarat, Rajasthan	Also known as Wadad or Waged, Wadhiar. Average milk yield 1738 kg per lactation.
Lakhimi	Assom	Average milk yield 359 kg per lactation, ranging from 325 to 375 kg with average milk fat 5.3%, ranging from 4.3 to 6.3%.
Nari	Gujarat, Rajasthan	Also known as Sirohi. Average milk yield 1647 kg per lactation, ranging from 1118 to 2222 kg with average milk fat 4.64%, ranging from 3.1 to 8.3%.
Ongole	Andhra Pradesh	Also known as Nellore. Average milk yield 1000 kg with milk fat 3.79%. Age at first calving 38-45 months with an intercalving period of 470 days. The breed is known for hardiness, disease resistance and capacity to thrive on scanty resources.
Purnea	Bihar	Average milk yield 609 kg per lactation, ranging from 452 to 785 kg with average milk fat 4.22%, ranging from 4 to 4.5 %.
Tharparkar	Gujarat, Rajasthan	Also known as White Sindhi, Grey Sindhi. Average milk yield 1749 kg per lactation, ranging from 913 to 2147 kg. Better heat tolerance and disease resistance.
Vechur	Kerala	Average milk yield 561 kg per lactation with milk fat ranging from 4.7 to 5.8%.

Source: http://14.139.252.116/agris/breed.aspx

https://vikaspedia.in/agriculture/livestock/cattle-buffalo/breeds-of-cattle-buffalo

2.2.2.2 Exotic cattle breeds

The exotic cattle breeds have got certain distinctive characteristics in comparison with indigenous cattle. They are high milk producers. Though high producers, they cannot withstand high temperature. Many exotic cattle breeds were brought to India for cross-breeding and upgrading purpose to improve milk production potentiality in indigenous cattle. The important exotic cattle breeds are Holstein-Frisian, Jersey, Brown-Swiss, Guernsey, Ayrshire, etc.

Table 2.2. Important exotic breeds of cows

Breed	Origin	Productivity
Holstein- Friesian	Holland	Milk yield 6000 kg/lactation with 3 to 3.5% fat. Age of maturity 12 months. Age at first calving 24 to 45 months. Calving interval 13 months.
Jersey	Jersey Island	Milk yield 4000 to 5000 kg/lactation with 4 to 4.5% fat. Age at first calving 25 to 26 months. Calving interval 12 months.
Brown Swiss	USA	Milk yield 5000 kg/lactation with 5% fat. Age at first calving 27 months. Calving interval 12 months.
Ayrshire	Scotland	Milk yield 4600 kg/lactation with 3.5-4% fat. Age at first calving 26 to 27 months. Calving interval 13 months.
Guernsey	Guernsey Island	Milk yield 4000 to 5000 kg/lactation. Age at first calving 26 to 27 months.

2.2.2.3 Indigenous buffalo breeds

Buffalo have significant potential for meeting India's dairy demand in a sustainable way. Buffalo are more adaptable to India's increasingly dry and hot climate. They also convert the low-quality indigenous grasses into milk more efficiently than dairy cows, improving productivity and reducing methane emissions. Consumers are increasingly choosing buffalo over dairy cow milk because they prefer the higher fat content of buffalo milk, while farmers prefer the higher returns they receive from buffalo milk compared to cow milk.

Table 2.3. Important buffalo breed for milk production

Breed	Breeding tract	Characteristics and yield potential
Mehsana	Gujarat	Milk yield 1800-2700 kg/lactation with 6.5-8.1% fat. Age at first calving 43 months. Intercalving period 450 to 550 days.
Murrah	Haryana, Delhi	Milk yield 1500-2500 kg/lactation with 7% fat. Age at first calving 42-52 months. Intercalving period 334-537 days. Used for the upgrading of inferior local buffalo.
Nili Ravi	Punjab	Milk yield 2000 kg/lactation with 6.5% fat. Age at first calving 45-50 months. Intercalving period 445-525 days.

Bhadawari	Uttar Pradesh, Madhya Pradesh	Milk yield 1100 to 1300 kg/lactation with 7-13% fat. Age at first calving 50-52 months. Intercalving period 454 days.
Jaffarabadi	Gujarat	Milk yield 1000 to 1200 kg/lactation with 9-10% fat. Age at first calving 40-45 months. Intercalving period 447 days.
Surti	Gujarat	Milk yield 1300-1400 kg/lactation with 8-12% fat. Age at first calving 36-38 months. Intercalving period 461 days.

Source: http://14.139.252.116/agris/breed.aspx
https://vikaspedia.in/agriculture/livestock/cattle-buffalo/breeds-of-cattle-buffalo

2.2.3 Housing Requirements

Adequate housing to dairy animals is aimed at increased milk production, better labour utilisation, better health of animals, disease control, better care and management of animals. Several types of housing are available. The farmer has to select the suitable housing system keeping in view the local environment and economic status. Minimum investment should be put towards housing of animals by utilising the locally available materials for construction of roof, floor and walls without compromising the comfort of animals.

2.2.3.1 Systems of housing

Housing systems available for dairy animals are conventional barn system and loose housing. Loose housing system is more suitable to Indian conditions. Another type of housing is free range system. It comprises of leaving the animals free in a large estate. Free ranges system or ranches indicate a type of stock management rather than a type of housing. The area is generally a natural or cultivated pasture land with watering points and shelter located at convenient places. This type of farming is suited to beef cattle.

Conventional barns

Cattle are more protected from adverse climatic condition in conventional barns. However, the conventional dairy barns are becoming less popular as they are comparatively costly. Animals in this system are confined on a platform and secured at neck by neck chains or ropes. The cows are fed and milked in the barn. The barns are completely roofed and walls are also complete with windows.

Advantages

1. Animals are less exposed to harsh weather conditions.
2. The animals can be kept cleaner and disease control is easier.
3. There is no need to have a separate milking parlour.

Disadvantages

1. The construction cost of shed is more when compared to loose housing system.
2. This is not suitable to hot and humid climates.
3. There is no scope for future expansion of shed.

Loose housing system

This system comprises of keeping animals loose in an open paddock or pasture throughout the day and night except at milking time. The open paddock is provided with shelter along one side under which the animals stay when it is hot or cold or during rains. A common watering tank and common fodder manger is provided within the shed. Concentrates are fed at time of milking in a separate milking barn. The open paddock is enclosed by half-walls and wooden or plain wire fences of suitable height. This type of housing is suitable to most parts of the country except in temperate Himalayan region and heavy rainfall areas. A milking barn or parlour is always to be constructed in which cows are milked at milking time.

Advantages

1. The loose houses are cheaper to construct, easier to expand and flexible in utility.
2. Feeding and management of stock is easier because of common feeding and watering arrangement.
3. Animals are more comfortable as they move about freely.
4. At least 10 to 15% more stock than the standard can be accommodated for shorter periods without unduly affecting their performance.
5. Detection of heat in animals is easier.
6. Animals also get sufficient exercise which is extremely important for better health.

Disadvantages

1. More floor space is required when compared to conventional barn system.
2. Competition is more for feed and fodder among animals.
3. Individual animal attention is not possible.
4. Provision for a separate milking barn is required.
5. This is not suitable in heavy rainfall and temperate areas.

2.2.3.2 Buildings for dairy animals

The buildings required for dairy animals may be of essential types and ancillary types.

Essential buildings

These buildings include milking barn, shed for milch/dry cows, maternity pens, calf sheds, young stock shed, bull shed and sick animal shed.

Milking barn: It is fully covered barn in which milch animals are milked and located at central place with all other buildings around it. The dimension of standing space depends on size of animal and normally it is 1.5-1.7 m × 1.1-1.2 m. The width of central passage should be 1.5 m × 1.8 m. There should be two continuous feed mangers on either side of shed with a 0.75 m wide feeding alley beyond each manger. A shallow U shaped drain about 20 cm wide is located on either side of central passage. The roof of shed should be gabled. The eaves of roof should be at least 50 cm away from side walls. Large open spaces may be left on side walls.

Milch/dry animal sheds: The sheds are for housing milch cows and dry cows separately. These are simple sheds comprising a closed area and adjoining open paddock. The covered area should be preferably concrete.

Maternity pens: Pregnant animals are transferred to maternity pens two to three weeks before date of calving. The number of maternity pens required is about 5% of number of breedable stock. The dimensions are 3 m × 4 m for covered area and another 3 m × 4 m for open paddock.

Calf sheds: The calf sheds are constructed nearer to the milking barn. The dimensions of calf sheds depend on the number of calves to be housed. If large number of calves are present then calves of different age groups should be housed separately for better feeding and management.

Young stock shed: Calves from six months of age to one year of age are to be housed separately from suckling calves. Generally all male calves above six months of age are disposed off.

Ancillary buildings

Feed stores: It is required for storing concentrates. There should be one concentrate-cum-feed mixing room at a distant place and a small feed ration room near the milking barn. The feed room must be damp-free and rodent proof. The size of the room is based on the assumption that 0.2 m storage space is required for each adult unit.

Silos: Under Indian conditions trench silos are convenient. It is constructed by excavation in the hard ground and floor and sides are lined with concrete or brick which are seepage proof.

Hay/straw shed: Sheds with walls on three sides are better for storing of hay/ straw. The shed should be away from animal sheds because of fire hazards.

2.2.3.3 Floor space requirements

Open or paddock areas of sufficient space is provided to cattle and buffaloes in warm regions since these animals lie down during the night. Maximum number of animals housed per pen also determines the comfort the animal gets.

Table 2.4. Floor space requirements for different types of animals under loose housing system.

Types of animal	Floor space requirement (m^2/animal)		Maximum number of animals/pen	Height of the shed (cm)
	Covered area	Open area		
Bulls	12	24	1	1.75 m in medium and heavy rainfall and 2.2 m in dry area
Cows	3.5	7	50	
Buffaloes	4	8	50	
Young-calves	1	2	30	

Feeding and watering space requirements

There should be a free access and sufficient feeding or watering space for the animals. The animals should be at comfortable position and with no competition while in activity. Feed-trough or manger must accommodate all the animals at a time. Water-trough must accommodate at least 10% of the stock.

Table 2.5. Feeding and watering space requirements

Type of animal	Space/animal (m)	Total manger length in a pen for 100 animals (m)	Total water tank length in a pen for 100 animals (m)
Adult cattle and buffaloes	0.60-0.75	60-75	6.0-7.5
Calves	0.40-0.50	40-50	4.0-5.0

2.2.4 Feed Management

Feed alone constitute 60% of the production cost of milk. Hence, feeding management plays a vital role in farm economy. The nutrient requirement should be determined for maintenance as well as for milk production and to meet the fat percentage in milk and gestation. Based on the nutrient requirement ration should be computed. In general the dry matter from roughage should not exceed 2% of cow's live weight nor should it be less than 1%.

Table 2.6. Feeding schedule for cows based on a thump-rule

Stage of lactation	**Quantity of green grass to be given (kg) for animal weighing**			**Concentrate (kg)**
	250 kg	**300 kg**	**350 kg**	
Dry cow	25	30	35	• For non-pregnant cows no concentrate is required. • Pregnant cows should be fed additional quantity of 1.5 kg of concentrate from 7th month of gestation. • In case of dry cow, allowance up to 1 kg concentrate can be given if the condition of cow is poor or the fodder quality is inferior.
Milch cow	25	30	35	• 1.0 kg for every 2.5 kg of milk of average 4% fat, in case of buffalo 1.0 kg for every 2.0 kg of milk produced.

Tips for feeding dairy cattle

- Concentrate must be fed individually according to production requirements.
- Good quality roughage saves concentrates. Approximately 20 kg of grasses or 6-8 kg legume fodder can replace 1 kg of concentrate mixture in terms of protein content.
- 1 kg straw can replace 4-5 kg of grass on dry matter basis. In this case the deficiency of protein and other nutrients should be compensated by a suitable concentrate mixture.
- Regularity in feeding should be followed. Concentrate mixture can be fed at or preferably before milking, half in the morning and the other half in the evening, before the two milkings.
- Half the roughage ration can be fed in the forenoon after watering and cleaning the animals. The other half is fed in the evening, after milking and watering. High yielding animals may be fed three times a day (both roughage and concentrate). Increasing the frequency of concentrate feeding will help maintain normal rumen motility and optimum milk fat levels.
- Over-feeding concentrates may result in off-feed and indigestion.
- Abrupt change in the feed should be avoided.
- Grains should be ground to medium degree of fineness before being fed to cattle.

- Long and thick-stemmed fodders such as napier may be chopped and fed.
- Highly moist and tender grasses may be wilted or mixed with straw before feeding. Legume fodders may be mixed with straw or other grasses to prevent the occurrence of bloat and indigestion.
- Silage and other feeds, which may impart flavour to milk, may be fed after milking. Concentrate mixture in the form of mash may be moistened with water and fed immediately. Pellets can be fed as such.
- All feeds must be stored properly in well-ventilated and dry places. Mouldy or otherwise damaged feed should not be fed.
- For high yielding animals, the optimum concentrate roughage ratio on dry matter basis should be 60:40.

Table 2.7. Feeding allowances for dairy cattle and buffalo

Type of cattle	Stage of the cattle	Green fodder (kg/day/ animal)	Dry fodder (kg/day/ animal)	Concentrates (kg/day/ animal)
Cow (average weight 250 kg)	Milk yield 5 litres/day	15.0	5.0	2.0
	Milk yield 5 to 10 litres/day	17.5	5.5	3.0
	Milk yield 10 to 15 litres/day	20.0	6.0	4.0
	Cow in gestation	15.0	5.0	1.5
Buffalo (average weight 400 kg)	Milk yield 5 litres/day	15.0	5.0	2.5
	Milk yield 5 to 10 litres/day	20.0	6.0	4.0
	Milk yield more than 10 litres/day	25.0	7.0	5.0

2.2.5 Breeding Aspects of Dairy Animals

Animals having body weight below 250 kg are not suitable for breeding. Cows remain in milk for about 300 days. A cow does not require more than 6-8 weeks of dry period. Cows should be bred during the second or third month after calving. If artificially inseminated, it is better to inseminate for 3 days continuously to have high probability to conceive. If oestrus signs are observed in morning, breeding is done at evening. If oestrus signs are observed in evening, breeding is done at next day morning. Pregnancy diagnosis should be done after 45-60 days of breeding by qualified veterinarian. This facilitates optimal feeding and care of pregnant animals in positive animals. This also provides clear way to breed the animal in next oestrus in negative animals. Gestation period varies with individual cows and breeds, normally it is about 280 days. In case of buffaloes, lactation period lasts for 7-9 months. Gestation period is about 310 days.

2.2.5.1 Infertility management

Infertility in cattle accounts for major economic losses in dairy farming and dairy industry in India. Maintaining an infertile animal is an economic burden. In cattle, nearly 10-30% of lactations may be affected by infertility and reproductive disorders. The causes of infertility are many and can be complex. Infertility or failure to conceive and give birth to a young one can be due to malnutrition, infections, congenital defects, management errors and ovulatory or hormonal imbalances in the female. To attain good fertility or high calving rate both the male and female animals should be well fed and free from diseases.

Both cows and buffaloes have the sexual cycle (oestrus) once in 18-21 days for 18-24 hours. But in buffaloes, the cycle is silent posing a big problem to the farmers. The farmers should closely monitor the animals 4-5 times from early morning to late night. Farmers who maintain good records and spend more time watching the animals obtain better results.

2.2.5.2 Tips to avoid infertility

1. Animals that do not show oestrus or do not come to cycle should be checked and treated.
2. Deworming once in 6 months should be done for worm infestations to maintain the health status of the animals. A small investment in periodic deworming can bring greater gains in dairying.
3. Cattle should be fed with a well balanced diet with energy, protein, minerals and vitamin supplements. This helps in increased conception rate, healthy pregnancy, safe parturition, low incidence of infections and a healthy calf.
4. Care of young female calves with good nutrition helps them to attain puberty in time with an optimum body weight of 250 kg, suitable for breeding and thereby better conception.
5. Feeding adequate quantity of green fodder during pregnancy will avoid blindness in newborn calves and retention of placenta after birth.
6. In natural service, breeding history of the bull is very important to avoid congenital defects and infections.
7. Infections of the uterus can be largely avoided by having cows served and calved under hygienic conditions.
8. Unwarranted stress and transportation should be avoided during the last stages of pregnancy.

9. The pregnant animal should be housed away from the general herd for better feeding management and parturition care.
10. Pregnant animals should be drained of their milk two months before delivery and given adequate nutrition and exercise. This helps in improving the health of the mother, delivery of a healthy calf with average birth weight, low incidence of diseases and early return of sexual cycle.

2.2.6 Health Management

Table 2.8. Vaccination schedule for dairy animals

Name of the vaccine	Age of first vaccination	Period of booster vaccination	Period for next vaccination
Hemorrhagic septicaemia (HS)	15 days	3- 4 weeks after first vaccination	Every year before rainy season
Anthrax	3 months	-	Every year before rainy season
Black quarter (BQ)	15 days after the Anthrax vaccination	-	Every year before rainy season
Foot and mouth disease (FMD)	6 months	4 months after first vaccination	Repeat every year in October/ November
Rinderpest	6 months	-	Repeat after 3/4 years

In addition to the vaccinations, the periodic health check-up of the dairy animals should be done. Animals should be treated for both ectoparasites (ticks, mites, etc.) and endoparasites (tape worm, round worm, etc.) at regular interval.

2.3 GOAT AND SHEEP

Small ruminants such as goat and sheep play an important role in Indian economy and source of livelihood and employment to millions of rural households especially the landless, marginal and small farmers. They also exploited commercially as the investment is very minimal due to low cost of shed and animals. The feed and fodder cost is low due to their dependence on variety of leaves and almost no competition with humans for feed. The financial return is also high and time required for realising the profit is very less.

The goat and sheep are mainly dependent on common property resources (CPRs) for meeting their feed and fodder requirements. The CPRs comprise barren and uncultivable lands, cultivable wastes, permanent pastures and other grazing land and land under miscellaneous trees, crops and other fallow land. The rural poor who cannot afford to maintain a cow or a buffalo find goat/sheep as the best alternative source of supplementary income and milk. Unlike a cow or buffalo, a few goats can be maintained easily and can be easily sold at the time of need.

2.3.1 Benefits of Goat Farming

1. The initial investment needed for goat farming is low.
2. Due to small body size and docile nature, housing requirements and managerial problems with goats are less.
3. Goats are friendly animals and enjoy being with the people.
4. Goats are prolific breeders and achieve sexual maturity at the age of 10-12 months. Gestation period in goats is short and at the age of 16-17 months it starts giving milk. Twinning is very common and triplets and quadruplets are not rare.
5. In drought prone areas risk of goat farming is very much less as compared to other livestock species.
6. Unlike large animals in commercial farm conditions both male and female goats have equal value.
7. Goats are ideal for mixed species grazing. The animal can thrive well on wide variety of thorny bushes, weeds, crop residues, agricultural by-products unsuitable for human consumption.
8. Under proper management, goats can improve and maintain grazing land and reduce bush encroachment without causing harm to the environment.
9. No religious taboo against goat slaughter and meat consumption prevalent in the country.
10. Slaughter and dressing operation and meat disposal can be carried without much environmental problems.
11. The goat meat is more lean (low cholesterol) and relatively good for people who prefer low energy diet and sometimes goat meat is preferred over sheep meat because of its chewability.
12. Goat milk is easy to digest than cow milk because of small fat globules and is naturally homogenised. Goat milk is said to play a role in improving appetite and digestive efficiency.
13. Goat milk is non-allergic as compared to cow milk and it has antifungal and antibacterial properties and can be used for treating urogenital diseases of fungal origin.
14. Goats are 2.5 times more economical than sheep on free range grazing under semiarid conditions.
15. Goat creates employment to the rural poor besides effectively utilising unpaid family labour. There is ample scope for establishing cottage industries based on goat meat and milk products and value addition to skin and fibre.

16. Goat is termed as walking refrigerator for the storage of milk and can be milked number of times in a day.

2.3.2 Benefits of Sheep Farming

1. Sheep do not need expensive housing facility and on the other hand require less labour than other kinds of livestock.
2. The foundation stock is relatively cheap and the flock can be multiplied rapidly.
3. Sheep are economical converter of grass into meat and wool.
4. Sheep will eat varied kinds of plants compared to other kind of livestock. This makes them excellent weed destroyer.
5. Unlike goats, sheep hardly damage any tree.
6. The production of wool, meat and manure provides three different sources of income to the shepherd.
7. The structure of their lips helps them to clean grains lost at harvest time and thus convert waste feed into profitable products.
8. Mutton is one kind of meat towards which there is no prejudice by any community in India and further development of superior breeds for mutton production will have a great scope in the developing economy of India.

2.3.3 Breeds and Breeding Management

India has vast genetic resources of small ruminants of which 26 breeds of goat and 42 breeds of sheep are well documented. These breeds are adapted to various agroclimatic zones of the country. The important indigenous goat breed include Jamunapari, Beetal, Barbari, Tellicherry, Sirohi, Osmanabadi, Black Bengal, Bidri, Ganjam, Gohilwadi, Jakhrana, Kahmi, Konkan Kanyal, Kutchi, Mehsana and Surti. Exotic goat breeds include Saanen, Alpine, Anglo-Nubian, Toggenberg, Angora and Boer.

Among the indigenous sheep breeds Chokla, Magra, Nali, Pugal, Marwari, Malpura, Sonadi, Jaisalmeri, Patanwadi, Muzzaffarnagri, Jalauni, Hisserdale, Mecheri, Vembur, Neelagiri, Trichy black, Deccani, Nellore, Mandya, Balangir, Bellary, Chokla, Kajali, Kendrapada, Nilgiri, Shahbadi and Tibetan are popular. Important exotic sheep breeds are Dorset, Suffolk, Merino, Rambouillet, Cheviot and Southdown.

Goat matures in about 6-7 months. Breeding is allowed for buck at one year and doe after 10 months of age. The average gestation period is between 145-155 days. It gives birth to 1-3 kids at a time. Kids are weaned at 7-8 weeks and then mother can be allowed for mating. Castration of kids should be done before 12 weeks for optimisation of growth and removing the musky odour of rams or bucks. In addition, if the hooves are growing or the animals are not doing enough walking then hoof trimming should be done in every 6-8 weeks.

One ram is maintained for 40-50 ewes. Ewes do not come in heat at regular intervals throughout the year, but are seasonal. Duration of the heat period is one to three days and ewes remain in heat for 21-39 hours. Optimum time for service is towards the end of heat period. Average heat interval is 18 days during the breeding season. Gestation period is 142-152 days. A ram is in full vigour for breeding at 2.5-5 years of age. Ewes are ready for breeding at two years of age. In every 2-3 years, the bucks or rams of an area should be interchanged with the bucks or rams from another area to prevent slow growth by inbreeding depression. As sheep are seasonal breeders and mostly breed during September to December, they may be given flushing ration, and special care must be taken.

2.3.4 Housing of Goats and Sheep

Simple shade with low cost housing materials is enough for sheep and goat for their optimum production efficiency. The optimal climatic environments for goat and sheep shelter are air temperature of 13-27°C, relative humidity of 60-70%, wind velocity of 5-8 km/hour and a medium level of solar radiation. The sheds should be constructed in an elevated area to prevent water stagnation in high rainfall areas. Floors of the shed should be firm and should have the capacity to absorb water. Sheds with mud floor are suitable for most of parts of the country except where high rainfall is observed. The floors should be constructed in such a way that it should be easily cleaned. For a comfortable house east-west orientation with generous provision for ventilation to dry the floor will be suitable one.

If the animals are kept in open area, a covered area and run space is required. The run space should be covered by chain links. The covered area is used for sheltering during night and adverse climatic conditions. Separate structures are required for sick animals and young ones. Regular cleaning of house and fortnightly application of lime on floor is essential for disinfection of shed. The manure pit should be constructed at a distance from the shed. The surrounding of shed must be free from weeds and stagnant water. Thatched roof is best suited one for its cheaper cost and durability. However, corrugated asbestos sheets can also be used for organised farms to minimise the recurring costs and longer durability. The provision of optimum floor space, which includes closed space and open space is highly essential for optimum growth of animals.

Table 2.9. Recommended space requirements for goats and sheep

Age groups	Covered space (m^2)	Open space (m^2)
Up to 3 months	0.20-0.25	0.40-0.50
3 months to 6 months	0.50-0.75	1.00-1.50
6 months to 12 months	0.75-1.00	1.50-2.00
Adult animal	1.50	3.00
Male, pregnant or lactating ewe/doe	1.50-2.00	3.00-4.00

When the animals are taken for grazing during the day time and sheltered only during night the covered space will be enough. When the animals are housed intensively the pen and run system of housing is suitable. There is no restriction for the length of the shelter, however breadth of shed should not exceed 12 m and optimum breadth of shelter is 8 m. Height of eaves should be 2.5 m and height at ridge should be 3.5 m. The height of chain link used for open space should be 1.2 m. The length of the overhang should be 0.75-1.0 m.

Table 2.10. Feeding and watering space requirement for goats and sheep

Type of animal	Space per animal (cm)	Width of manger/ water trough (cm)	Depth of manger/ water trough (cm)	Height of inner wall of manger/ water trough (cm)
Sheep and goat	40-50	50	30	35
Kid/lamb	30-35	50	20	25

2.3.5 Feed Management

The new-born kid/lamb should be offered sufficient amount of colostrum immediately after birth and then sufficient of milk. The young ones will nibble after 15 days and slowly eat green roughages along with milk. The creep feed may be started from one month of age and up to 2-3 months of age. The main purpose of creep feeding is to give more nutrients for their rapid growth. The general quantity to be given to the kids/lambs is 50-100 g/animal/day. This should contain 22% protein. Composition of ideal creep feed is maize 40%, ground nut cake 30%, wheat bran 10%, deoiled rice bran 12%, molasses 5%, mineral mixture 2% and salt 1% fortified with vitamins A, B2 and D3. Antibiotics like oxytetracycline or chlortetracycline may be mixed at the rate of 15 to 25 mg/kg of feed.

Table 2.11. Feeding schedule for a kid/lamb from birth to 90 days

Age of kids/lambs	Mother's milk or cow milk (ml)	Creep feed (g)	Forage, green fodder (g/day)
1-3 days	Colostrum 300 ml, 3 feedings		
4-14days	350 ml, 3 feedings		
15-30 days	350 ml, 3 feedings	A little	A little
31-60 days	400 ml, 2 feedings	100-150	Free choice
61-90 days	200 ml, 2 feedings	200-250	Free choice

After 90 days, adequate grazing in the pasture and 50-200 g/animal/day concentrate mixture depending upon the age, pregnancy and lactation is given. Dry fodder is also given during night in summer months and during rainy days. In last one month of pregnancy, the foetal growth is the maximum and lack of enough energy in the feed can cause pregnancy toxaemia in ewes. In addition to grazing, the pregnant animals should be fed with concentrate mixture of 150-250 g/animal/day till 3-4 days before parturition. It is usually preferable to feed lightly on the day of parturition, but allow plenty of clean, cool water. Similarly, during breeding time concentrate must be fed for optimising reproduction. Separate feeders and water troughs should be placed for concentrate feeds, green fodders and water. Feeding and watering should follow the same schedule, everyday. If feed is changed suddenly animal may not feed properly, so new feed is introduced slowly and in small amounts till animal gets used to it.

Clean and wholesome drinking water should be available in sufficient quantity. If not fed any concentrate mixture then 5-10 g of mineral-vitamin mixture should be supplemented, particularly before mating, last part of pregnancy and after kidding for 2-3 months.

Goats are essentially browsers and eat plants, which any other animal cannot touch. They prefer tree fodder. So 40-50% of green fodder should contain tree leaf fodder and rest with other grasses. Leaves or branches should be kept in hanging condition for goats. Tree crops like subabul, ber, agasti, siris can be grown on the pond dyke or farm boundary to feed the leaves of those trees to the goat. Feed consumption of goat is 3% of its body weight/day. By considering the availability of roughages in farming system in different season the rest of the nutritional need of the goats can be met by feeding rice bran, what bran, etc.

Table 2.12. Rate of feeding concentrate per day

Body weight (kg)	When legume fodder is available (g)	When legume fodder is not available (g)
Up to 12	25-50	200-300
12-15	50	300
15-25	100	400
25-35	150	600

2.3.6 Parasites and Disease Management

Parasites are one of most important constraints which decrease the body weight gain and productivity of animals. So, the small ruminants should be dewormed on a regular basis, i.e. every 8-12 weeks. In goat and sheep, the major losses in productivity and lives are due to diseases like enterotoxaemia (sudden death in healthy animals after returning from grazing), peste des petits ruminants (PPR, also known as 'goat plague', is a viral disease with signs of fever, loss of appetite, stomatitis, gastroenteritis, profuse serous nasal discharge accompanied by sneezing, coughing and pneumonitis), neonatal death, foot and mouth disease and pox (with pox lesions throughout the skin and mucous membranes). These diseases have high mortality, especially the first two diseases. Once started, there is not much one can do and the loss is very high. So vaccinating the animals should be done to get the maximum from small ruminant production. Weak, debilitated, pregnant and infested animals should not be vaccinated. They should be dewormed prior to vaccination. Booster dose must be given whenever available as it strengthens the protection from disease. Generally vaccination against enterotoxaemia, PPR and pox are given after 4 months of age for annual protection; but foot and mouth disease (FMD) vaccine has to be given twice a year. Vaccination is a must for endemic regions which saves many lives and protects from economic loss. In case of disease outbreak, plan for cleaning and disinfecting house and equipments is needed. Neonatal mortality can be prevented by improving the level of nutrition in advanced stages of pregnancy, ensuring hygienic condition in the kidding sheds, providing proper bedding, and ensuring early feeding of colostrums.

Table 2.13. Vaccination schedule for goats and sheep

Name of disease	Primary vaccination	Regular vaccination
Anthrax	At the age of 6 month for kid or lamb	At the age of 6 month for kid or lamb
Haemorrhagic Septicaemia	At the age of 6 month for kid or lamb	Once annually before monsoon
Enterotoxaemia	At the age of 4 month for kid or lamb (if dam is vaccinated). At the age of 1st week for kid or lamb (if dam is not vaccinated)	Before monsoon (preferably in May). Booster vaccination after 15 days of first vaccination.
Black Quarter	At the age of 6 month for kid or lamb	Once annually (before monsoon)
Peste des petits ruminants	At the age of 3 month for kid or lamb and above	Once in three years
Foot & mouth disease	At the age of 4 month for kid or lamb and above	Twice in a year (September & March)
Goat Pox	At the age of 3 month and above	Once annually (December)

2.4 PIGGERY

Pig farming provides the livelihood to many rural resource-poor farmers. In India pig rearing is very popular amongst the tribal backward and weaker section of the society, especially in North Eastern region. It gives direct cash return through sale of livestock and indirect return in terms of manure and fuel. Pig grows fast and is a prolific breeder, farrowing 10-12 piglets at a time. A boar weighs 90 kg in just 5-6 months.

2.4.1 Breed Selection and Breeding Management

The exotic pig breeds mainly comprise Hampshire, Large White Yorkshire, Duroc, Landrace and Tamworth while some of the popular indigenous breeds include Ghungroo, Niang Megha, Ankamali and Agonda Goan. Large White Yorkshire is the most extensively used exotic breed in India. Body colour is solid white with occasional black pigmented spots. This is the best breed for the purpose of cross breeding. Mature boar weighs 300-400 kg while sow (female pig) weighs 230-320 kg. Landrace is another exotic breed which is excellent for crossbreeding. Mature boar weighs 270-360 kg and mature sow weighs 200-320 kg.

Reproduction is the main component limiting the productive efficiency of pig. Successful reproduction is the outcome of a series of closely linked events. The gilt (female pig under the age of one year and first farrowing) must grow rapidly to attain sexual maturity, initiate oestrous cycle, ovulate and be mated by a fertile boar or artificially inseminated. A gilt becomes sexually mature between 8-10 months depending on the breed and nutrition level. The length of the oestrous cycle averages approximately 21 days. It is always better to leave one or two oestrous cycle in case of the gilt and breeding may be done during the third cycle depending on the physical condition of the gilt. The best time for AI of sow/gilt is 15-24 hours after the onset of oestrus. If the female does not conceive, it will repeat its heat symptoms after 21 days. Those who do not repeat are presumed to be pregnant which, however, needs to be confirmed by a veterinarian.

Pregnant sow should be shifted to a clean farrowing house before 3 weeks of farrowing. Clean and dry bedding material preferably of dry paddy straw/hay has to be provided in the pen. The pregnant animal should be fed individually. New born piglets are active and within two minutes each piglet reaches a teat and attempt to suck milk. Sometimes respiration is delayed in newborn piglets. To stimulate the respiratory activity of the piglet the mucous should be removed from nose and mouth. The weaker piglets should be assisted to the teat so that they can suckle the colostrums. Care should be taken to avoid crushing of newborn piglets during and after farrowing.

2.4.2 Housing of Pigs

Pigs can be kept under open-air system and indoor system. A combination of both may be followed. It is easy to manage animals in indoor system compared to open-air system. However, in this system more area is required and possibilities of contamination of diseases are more than indoor system. Each animal of different categories requires a minimum floor space for housing.

Table 2.15. The floor space for different categories of pigs

Category	Covered area/pig (m^2)	Open space/pig (m^2)
Weaner (up to 20 kg body weight)	1.0-1.5	1.5-2.0
Grower	1.2-2.0	2.0-3.0
Dry sow	2.0-3.0	3.0-4.5
Lactating sow	6.5-9.5	6.5-9.5
Boar	3.5-4.5	4.5-6.5

Concrete floor with rough surface is recommended. The floor should be tilted to one side for easy cleaning and washing. The roof should be waterproof and good conductor of heat. The roof should be at least 8-10 feet high. The side walls of the sheds should be made up of brick and cement up to 4-5 feet high and remaining height may be fitted with iron wire netting. Feeding and watering trough are to be provided in each pig chamber. The feeding trough should be 15 inches wide and about 4 feet long with a partition in between for supplying water on one side and feed on the other side.

2.4.3 Feeding Management

Pigs are the most rapidly growing livestock and the most efficient animal in converting feed to meat. About 70-75% of the total production cost of the pig farm is due to the feed cost. It is therefore, very important to feed the animals with economical but balanced feed which will contain all the nutrient requirements for growth and to support the life.

The protein requirement in the ration for suckling pig before weaning is 22%, weaner 18-20% and the breeding boar and pregnant female 15%. The major minerals needed pig diets are calcium, phosphorus, iron, manganese, zinc, iodine and copper. Vitamins are also to be supplemented in the ration.

Table 2.16. Feed formula for different categories of pigs

Ingredients	Weaner (18-20% protein)	Growers (15-17% protein)		Gilt and sow (14-16% protein)
		12th weeks to market age	9-12 months	
Maize	55	58	60	50
Groundnut cake	17	15	8	13
Wheat bran	20	20	25	20
Rice polish	- -	- 10		
Fish meal/soya meal	6	5	5	5
Mineral mixture	1.5	1.5	1.5	1.5
Salt	0.5	0.5	0.5	0.5

The pigs can also be maintained with vegetable waste, kitchen waste consisting of cooked rice, vegetables, etc. and with brewery waste and other feed waste. A good feed ration should contain the required nutrient in right proportion as per the need of body weight along with adequate drinking water.

Table 2.17. Concentrate requirement on body weight basis

Particulars	Body weight (kg)				
	<10	10-20	20-30	30-60	>60
Protein level (%)	20	18	16	16	13
Concentrate feed kg /animal/day	0.5	1.5	1.5	2	2.5

2.4.4 Diseases and Health Care

The pigs are infected with a number of internal parasites, skin infections and other bacterial and viral diseases, which may result in poor growth and even death. In young pigs, infection with roundworms can cause diarrhoea and weight loss. The piglets should be dewormed regularly once in every three months. The pregnant sow should be dewormed before giving birth; otherwise the infection would pass to her litter. All the new-borns should be vaccinated against swine fever at the age of 2 months. Illness in pig is characterised by dullness, inappetite, disinclination to move, rough coat with loss of bloom, constipation or diarrhoea. Once these symptoms are observed a veterinarian should be consulted. Most of the diseases can be prevented by following strict hygienic measure and by timely vaccination of pigs. A separate house should be there to keep the animals suffering from contagious diseases.

2.5 RABBIT FARMING

The agricultural practice of breeding and raising domestic rabbits as livestock for their meat, fur, or wool is termed as cuniculture. Rabbit farming is a business with low investments and high profits. However, marketing is a big problem for commercial rabbit production in India. Rabbit meat has good demand in only some areas of the country. Thus, local markets should be expanded and efforts may be made to export the rabbit meat and fur to foreign countries.

2.5.1 Advantages of Rabbit Farming

1. Rabbits are highly prolific with some females producing 25 to 50 kits (young ones of rabbit) per year.
2. The meat of the rabbit is rich in poly-unsaturated fatty acids which come under the category of white meat.
3. Small groups of 50 rabbits can reared in the backyard of the house with kitchen waste as feed.
4. The initial investment cost for rabbit farming is low along with quick returns (about six months after the establishment of the farm). Income generation at quarterly interval makes the repayment easy.
5. They are the best producers of wool on per kilogram body weight basis and require 30% less digestible energy to produce 1 kg of wool as compared to sheep. Rabbit's wool is 6-8 times warmer than the sheep wool. It can be mixed with silk, polyester, rayon, nylon, sheep wool and other fibres to make good quality handlooms as well as hand knitted apparels. There is a high demand for fur obtained from Angora rabbits in India as well as in foreign countries due to which it can be a successful business venture.
6. Residual feed, together with rabbit manure is highly suitable for vermicomposting which provides excellent manure to be used as an organic fertiliser in agricultural fields.
7. Rabbits consume a large amount of forage from diverse origins and hence, can be reared on roughage with very less quantity of costly concentrate feed.

2.5.2 Breed Selection and Breeding Management

There are numerous breeds available throughout the world. Some of these breeds are highly productive and very suitable for farming in India. Rabbit breeds such as White Giant, Grey Giant, Flemish Giant, New Zealand White, New Zealand Red, Californian, Dutch and Soviet Chinchilla have been identified for our country.

Rabbits become suitable for breeding at 5 to 6 months of age. Male rabbits at one year of age should be used for breeding purpose. Healthy rabbits should be chosen for breeding with proper age and body weight. Proper care of breeding male rabbits as well as that of pregnant females must be taken.

2.5.3 Housing and Feeding Management

Rabbits can be raised in both the deep litter system and cage system. Good housing is necessary for keeping the rabbits free from weather conditions, rain, sun and various types of predators like dogs or cats, etc. A small shed can be made at the backyard.

Rabbits can eat and consume all types of grains, legumes and green fodders like *Lucerne*, *Sesbania*, *Desmanthus* and various types of kitchen wastes including carrots, cabbage leaves, and other vegetable wastes. In case of raising rabbits by feeding concentrates, for each kg body weight, about 40 g concentrate and 40 g green fodder should be fed daily. Along with fresh and nutritious food, the rabbits should be provided with sufficient amount of fresh and clean water according to their demand.

Generally, diseases are less in rabbits. Healthy rabbits are very active with a shiny hair coat. However, if they become dull or something goes wrong related to their health a veterinarian should be consulted.

2.6 CHICKEN FARMING

Poultry, defined as domesticated birds reared either for their meat, eggs, or the both, constitute a fascinating and diverse group of animals, which can be successfully integrated into many farming systems throughout the world. Poultry includes laying hens, meat chickens, ducks, turkeys, quails, guinea fowls, emu, pigeon, etc. The ability of these birds to convert many types of feed, such as residuals from agricultural activities, households and food processing industries, into animal products and protein sources is more efficient than most other animal species.

2.6.1 Breeds of Chicken

There are only four pure Indian breeds of chicken available. These are Aseel, Chittagong, Ghagus and Busra. Native chicken have undergone natural selection and are best adapted for the local climatic conditions. They can survive well on scavenging and leftover feed. Native hens are good brooders and have good mothering ability. But native chicken are low on productivity. There are several varieties of indigenous domestic fowl found in India which vary in their colour, size and shape. They are collectively called *desi* chicken.

Table 2.18. Indian breeds of chicken

Breed	Home tract
Ankaleshwar	Gujarat
Aseel	Chhattisgarh, Odisha, Telengana and Andhra Pradesh
Busra	Gujarat and Maharashtra
Chittagong	Meghalaya and Tripura
Daothigir , Miri	Assam
Ghagus	Karnataka, Telengana and Andhra Pradesh
Harringhata Black	West Bengal
Kadaknath	Madhya Pradesh
Danki, Kalasthi	Telengana and Andhra Pradesh
Kashmir Favorolla	Jammu and Kashmir
Nicobari	Andaman & Nicobar
Punjab Brown	Punjab and Haryana
Tellichery	Kerala
Mewari	Rajasthan

There is a wide variety of improved chicken breeds, developed for egg production, meat production, or for dual purpose. Many of these breeds are adaptable to a backyard setting. Improved varieties like Cauvery, Kalinga Brown, Nirbheek, Gramapriya, Blackrock and Vanaraja due to their moderate bodyweight and high egg yielding capacity are more preferred under semi-intensive farm conditions in rural areas. Some other improved breeds are Giriraja, Swarnadhara, Gramalakshmi, Gramasree, Srinidhi, Jharsim, Kamrupa, Pratapdhan, Krishibro, Shyama, Priya, Sonali , Devendra, Rainbro, etc.

The exotic breeds acclimatised to India include Rhode Island, Red Rhode Island, White Plymouth Rock, New Hampshire, Leghorn, White Leghorn, Minorca, Cornish, Sussex, Orpington, Dorking, California Grey, Australop, etc. Majority of the stocks used for egg production are crosses involving the strains or inbred lines of White Leghorn.

Advantages of indigenous breeds over exotic breeds

1. Acceptability of the coloured *desi* bird by the landless labourers or marginal farmers.
2. Use of broodiness for hatching the chicks.
3. Capability of self-defence from predators due to its alertness, light bodyweight, longer shank length, camouflagic characters and aggressiveness.
4. Can thrive well under adverse environments like poor housing, poor management and poor feeding.

5. Indigenous birds are comparatively disease resistant to protozoon and ecto-parasites.
6. Have better adaptability to extreme climatic conditions prevailing in the country.
7. They are comparatively hardier and need less health care than exotic birds.
8. The meat from native fowl has significantly higher amino acid contents (arginine and lysine) than meat from exotic birds and is widely preferred especially because of their pigmentation, taste, leanness and suitability for special dishes and often fetches higher prices.
9. The brown-shelled eggs of native fowl are rich in threonine and valine than farm eggs, have good flavour and fetch premium price.

2.6.2 Chicken Housing Management

Birds can be raised in both free range and indoor production systems. In case indoor production system, it is very crucial to manage the environment. Chickens need accurate management and environment for better production. Sufficient space should always be kept in the houses so that birds can live, grow and produce happily. Poultry house should never be overcrowded. A minimum 0.3-0.5 m^2 floor space is required per bird. The house should be well ventilated and sufficient flow of fresh air and light inside the house should be ensured. The chicken houses should be provided with wide windows and side curtains to ensure the entrance of fresh air to make the house cold during summer season. A pair of vent in the roof of the house will help to remove the hot air. A roof of at least six feet height from the floor is suitable for this purpose.

The house and equipment are to be cleaned on a regular basis and the house should be sterilised before bringing new chicks. Care must be taken for suitable drainage system inside the house for cleaning it properly. Predators and harmful animals should be kept away from the shelter. The main predators are rats, owls, hawks, stray dogs and cats. A quality coop is essential to backyard chicken production. Layers need nest boxes, one per 4-5 birds. Coops must provide protection from the weather. There should be a well-insulated area with a light bulb or heat lamp for the winter months as well as ventilation for fresh air. The housing system must have to be weather proof.

Light is also a very important element for chicken farming as birds are very sensitive to light. Light helps the birds to be productive, finding food and simulating them for reproduction. Besides light, the poultry birds also need dark period for keeping good health and producing melatonin hormone, which is very important for immune function. Birds require 8 hours of darkness period and 16 hours presence of light. Darkness is helpful for some fast growing broiler species, and helps them for reducing leg disorders and builds their body frame. However, young chicks require 24 hours light daily after hatching for finding food and water pot.

Usually the litter is used for covering the floor of housing system which may be made of concrete, wood or earth. Litter absorbs the moisture of closet and dilutes the manure. It also works as the bed for the poultry birds. Rice hulls, hay and straw and soft wood shavings are the common materials used as litter in chicken house floors. A good litter contains about 20 to 30% moisture and depth about 5 to 8 cm. High moisture in litter is very harmful for chicken health. Aluminium sulfate or hydrated lime may be used to reduce ammonia gas from the litter. After disposal of all birds the used litter from house floor should be removed. This litter can be used as good manure in the agricultural land for crop production. Dry poultry manure contains about 3.5% nitrogen, 2.0% phosphorus and 1.5% potassium.

2.6.3 Chicken Feed Management

Feeding is the major management concern in chicken farming as it accounts for 60-70% of total production cost. Birds require more than 40 nutrients which are available to them through water, proteins, carbohydrates, fats, minerals and vitamins. Carbohydrates and fats are the principal sources of energy. Fats are also the source of essential fatty acids, i.e. linoleic, linolenic and arachidonic acids. The requirement for protein is essentially the requirement for amino acids. Minerals (calcium, phosphorus, sodium. copper, iodine, iron, manganese and zinc) and vitamins play an important role in the regulation of several essential metabolic processes in the body. Any vitamin deficiency in poultry feed rations can hinder the natural growth of young poultry birds. Vitamin A comes from green feed ingredients, yellow corn and fish oils. It is essential to protect the poultry birds against colds and infections. Vitamin D helps to prevent leg weakness and rickets. It is found in synthetic products and also available in sea fishes. Vitamin D is a must added element in poultry feed. Vitamin B-complex is available in milk, green feed, liver, synthetic riboflavin, etc. It helps to increase the growth of chickens. It also helps to prevent curled-toe paralysis in young chickens. While preparing poultry feed, adequate amount of vitamin B-complex should be added in the poultry feed mixture.

Conventional poultry rations usually include many cereals like maize, rice, wheat, oat, barley; and a few cereal by-products such as wheat-bran or rice-polish, animal and vegetable protein sources like fish-meal, meat-meal, soybean-oil-meal, groundnut-cake, etc. The whole ration is fortified with adequate minerals and vitamins. A healthy laying hen diet should also contain crushed oyster shell for egg production, and grit for digestion. A critical part of a chicken's diet is continual access to clean, fresh water.

During the first day, feed may be sprinkled or provided in the trays for encouraging the new born chicks to pick up feed. From day two onwards feed is provided in trough type of feeders. As the chicks grow bigger suitable feeders are used. The feeders should be at proper height for the birds to eat properly. As chicks grow the feeder should be lifted up by adjusting their height to the back level of the bird. The level of feed in feeder has a direct correlation with feed wastage. As a thumb rule, 10% feed is wasted if the feeders are two thirds full compared to 3% wastage if they are half full and only one percent if they are one third full. Therefore, feed should be offered more frequently with small quantity at each time and helps to gain weight more uniformly. During first week, frequent feeding of small quantity should be practised to stimulate feed consumption.

Water may be provided using troughs, bell shaped drinkers and caps. With these drinkers 2 cm of water space per bird is recommended. Water should be provided before the chicks are released under the brooders. Height of the drinkers needs to be adjusted according to the chick height. One chick drinker is enough for 100 chicks up to 2 weeks of age and regular bell drinker is sufficient for 50 birds from third week onwards.

2.6.4 Chicken Disease Management

Various types of poultry diseases can cause serious loss in the poultry farming. However, serious illness is unlikely in a backyard flock, especially if the chickens are vaccinated. Antibiotics do not help viruses in birds. Proper nutrition, keeping the birds in clean environment and proper vaccination will help avoid most chicken diseases and symptoms. Preventative measures for most of the chicken diseases are always better than curative measures.

The birds must be vaccinated against Marek's and Ranikhet diseases. Birds should be dewormed starting from 7th week and repeated at 3-4 months intervals so as to give a total of 4 deworming. Piperazine compounds, albendazole, mebendazole, etc. can be used against roundworms. While medicating through drinking water, it should be done by mixing the required quantity of medicine in the quantity of water that chicks normally consume in 4 hours' time. One hundred chicks of 6 week-old consume approximately 6 litres water/day. Additional water should be given only when all the medicated water is consumed by the chicks.

Table 2.19. Vaccination schedule for poultry birds

Age (days)	Name of the vaccine	Dose	Route of administration
1	Marek's disease (HVT strain)	0.2 ml	Subcutaneous
4-7	New castle disease (Ranikhet disease) - F_1/ Lasota strain	One drop	Intraoccular/Intranasal (Eye/nasal)
14	Infectious bursal disease (Gumboro) - MB strain	One drop	Intraoccular/Intranasal (Eye/nasal)
35	New castle disease (Ranikhet disease) - R_2B strain	0.5 ml	Subcutaneous
42	Fowl pox	0.2 ml	Wing web prick

2.6.5 Backyard Chicken Farming

Most of the indigenous and improved chicken breeds are suitable for backyard rearing. It requires hardly any infrastructure setup. The birds are provided with lots of kitchen waste and food leftovers. They scavenge and eat termites, ants, earthworm and insects. During the dry seasons, the birds may additionally be provided with rice bran or wheat bran mixed with the rice starch and other vegetables. It has a very positive impact to improve the socioeconomic factors of the socioeconomically backward people. Backyard chicken can be taken up by every household as a subsidiary occupation, as a source of additional income or to cater to the egg and meat requirements of the family by taking up coloured bird units ranging from 10-20 birds per family in their backyards. Such units require very little hand feeding and can give a fairly handsome return with bare minimum night shelter.

Semi-intensive system may be adopted where the amount of free space available is limited, but it is necessary to allow 10-20 m^2 area per bird of outside run. In the semi-intensive system of rearing, feed and water are placed in the outside run or yard. The floor in the run can be cemented with addition of litter over the cemented floor or it could just be a sand floor. The run is fenced by using wire, bamboo or any other local material. The birds are driven to the pen during dusk, and they are well protected during the night. At daytime the birds are let out into the yard or run and stay there till the evening. The yard and the pen must be cleaned regularly once the birds are out of them. Under the semi-intensive system of rearing, birds are provided with feed and clean potable water. The feed could be a well balance mash or pellet, or it could be the remains of household waste, vegetable waste mixed with some cheap grains and by-products of grains.

2.7 DUCK FARMING

Ducks form about 10% of the total poultry population and contribute about 6-7% of total eggs produced in the country. Ducks are mostly concentrated in the eastern and southern states of the country.

Advantages of duck farming

1. Ducks lay more egg per bird per year than chicken.
2. The size of the duck egg is larger than hen egg by about 15 to 20 g.
3. Ducks require lesser attention and thrive well in scavenging conditions.
4. Ducks supplement their feed by foraging. They eat fallen grains in paddy fields, insects, snails, earthworms, small fishes and other aquatic materials.
5. From commercial point of view, ducks have a longer profitable life. They lay well even in second year.
6. Ducks do not require any elaborate houses like chicken.
7. Ducks are quite hardy, more easily brooded and more resistant to common avian diseases.
8. Marshy river side, wet land and barren moors upon which chicken or no other type of stock will flourish, are excellent quarters for duck farming.
9. Ducks lay 95-98% of their eggs in the morning before 9.00 AM. Thus it saves lot of time and labour of the grower for collecting eggs.
10. Ducks are suitable for integrated farming systems such as duck-cum-fish farming, duck farming with rice cultivation. In duck-cum-fish farming the droppings of ducks serve as feed for the fishes and no other feed or manuring of the pond is necessary for fishes (200-300 ducks per hectare of water area). Under integrated duck farming with rice cultivation, the ducks perform four essential functions, viz. intertillage as they search for food, their bills loosen up the soil around the rice plants, insect control and manuring.
11. Ducks are good exterminators of potato beetles, grasshoppers, snails and slugs. In areas plagued with liver flukes, ducks can help correct the problem. Ducks can be used to free the waterbodies from mosquito pupae and larvae.
12. Ducks are quite intelligent, can be tamed easily, and trained to go to ponds and come back in the evening of their own.

2.7.1 Duck Breeds

Among the egg laying breeds, 'Khaki Campbell' is the best producer. A duck lays about 300 eggs per year. Khaki Campbell ducks weigh about 2 to 2.2 kg, and drakes 2.2 to 2.4 kg. Egg size varies from 65 to 75 g. Other egg laying breeds are Nageswari, Indian Runner and Chara Chemballi.

'White Pekin' is the most popular duck for meat purpose. It is fast growing and has low feed consumption with fine quality of meat. It attains about 2.2 to 2.5 kg of body weight in 42 days of age. Other meat type breeds are China duck, Maskovi and Ruel Kagua.

2.7.2 Management of Stock

Brooding (0-4 weeks): The brooding period of 'Khaki Campbell' ducklings is 3 to 4 weeks. For meat type ducklings such as 'Pekin', brooding for 2 to 3 weeks is sufficient. A hover space of 90 to 100 cm^2 is required per duckling under the brooder. A temperature of 29 to 32° C is maintained during the first week. It is reduced by about 3° C per week till it reaches 24° C during the fourth week. Ducklings may be brooded in wire floor or litter. A wire floor space of 0.046 m^2 per bird or solid floor space of 0.093 m^2 per bird would be sufficient up to 3 weeks of age. Water in the drinkers should be 5 to 7.5 cm deep just sufficient to drink and not dip themselves.

Rearing (5-16 weeks): Ducklings may be reared in intensive, semi-intensive or range system. Under intensive system, a floor space of 0.279 m^2 per bird is allowed up to 16 weeks of age. Under semi-intensive system, a floor space of 0.186 to 0.279 m^2 per bird is allowed in night shelter and 0.929 to 1.394 m^2 as outside run per bird up to the age of 16 weeks. Usually ducklings are allowed to move to runs at the end of 3 to 4 weeks of age depending upon weather. Water in the drinkers should be 12.5 to 15 cm deep to allow minimum immersion of their heads. Partitions up to the height of 60-90 cm inside the pens and the outside runs are adequate for control. Under range system a flock of 1000 can be reared per 0.4 hectare.

Adult Stock (above 17 weeks of age): Under intensive system, a floor space of 0.371 to 0.465 m^2 per duck is essential, whereas in semi-intensive system, a floor space of 0.279 m^2 in the night shelter and 0.929 to 1.394 m^2 as outside run bird would be adequate. For wet mash feeding in a 'V' shaped feeder, 10 to 12.5 cm feeding space per duck is allowed but for dry mash or pellet feeding adlib in hoppers, a feeding space of 5 to 7.5 cm per duck would be sufficient. High egg laying strains of ducks come into production at 16 to 18 weeks of age. About 95 to 98% of eggs are laid by 9.00 AM. One nest box of size 30 cm × 30 cm × 45 cm to every three ducks is provided. In case of laying breeds a mating ratio of 1 drake to 6-7 ducks and in table breeds 1 drake to 4-5 ducks is allowed. Photoperiod of 14 to 16 hours per day is essential for optimum production.

While handling ducks, they should be caught by neck and not on the side of the body as this might lead to sudden death.

2.7.3 Duck Housing

Ducks do not require elaborate houses. Usually adult ducks do not require housing. If they are provided with some protection they will lay eggs better and damage less food. Ducks generally love darkness. The house should be well ventilated, dry and rat proof. The roof may be of shed type, gable or half round. The floor of the duck house may be build with dirt, concrete, wood or bamboo. Straw can be used inside the house as bed. It is very easy to clean and make dry of the floor, made with straw bed. Under semi-intensive system the house should have easy access to outside run as the ducks prefer to be outdoors during the day time and even during winter or rains. The run should gently slope away from the houses to provide drainage.

2.7.4 Duck Feeding and Management

Though duck is a water fowl and very fond of water, water for swimming is not essential at any stage of duck rearing. However, water in drinkers should be sufficiently deep to allow the immersion of their heads and not themselves. If they cannot do this, their eyes seem to get scaly and crusty and in extreme cases, blindness may follow. In addition, they also like to clean their bills periodically and wash them to clear off the feed.

Ducks may be grown on dry mash, a combination of dry and wet mash or pellets. Ducks prefer wet mash due to difficulties in swallowing dry mash. The pellet feeding, though slightly costly, has distinct advantages such as saving in amount of feed, minimum wastages, saving in labour, convenience and improvement in sanitary conditions. Ducks are good foragers. The use of range, pond or supplementary green feed, reduces the feed cost. Ducks should never have access to feed without water. During the first eight weeks, birds should always have access to feed, but later on they may be fed twice a day, i.e. first in the morning and then late afternoon. A 'Khaki Campbell' duck consumes about 12.5 kg of feed up to 20 weeks of age. Afterwards the consumption varies from 120 g and above per bird per day and depending upon the rate of production and availability of greens.

The concentrate feed should contain 16% protein for layer bird and 20% protein for broiler finisher bird. Starter ration (0-8 weeks) and grower ration should contain 22-24 and 20% protein respectively. Feed ingredients for 10 kg feeds for adult birds should have maize 4 kg, rice polish 3 kg, mustard oilcake 1 kg, mineral mixture 350 g and salt 250 g, fortified with vitamins and antibiotics. All ingredients should be free from aflatoxin.

2.7.5 Diseases Management

Ducks are resistant to common avian diseases.

Duck plague: Adult birds are mostly affected by virus disease. It is characterised by vascular damage with tissue haemorrhages and free blood in body cavities. The lumina of intestine and gizzard are filled with blood. There is no treatment for the disease. The birds can be protected by vaccine, which is given at the age of 8-12 weeks.

Duck cholera: It is an infectious disease, caused by bacterial organism *Pasteurella multocoda* in ducks over four weeks of age. There is loss of appetite, high body temperature, thirst, diarrhoea and sudden death. The disease can be prevented by vaccination of the birds with duck cholera vaccine, first at the age of 4 weeks and again at 18 weeks.

Parasites: Ducks are resistant to internal parasites. The infestation is prevalent only among those ducks which have access to stagnant water, over-crowded ponds and small streams. The parasites include flukes, tapeworms and roundworms. These cause decrease of nutrient assimilation by the bird and anaemia due to toxic material excreted by them, destroying the red cells. The external parasites include lice, mites, fleas and ticks. These cause irritation and annoyance leading to loss in egg production. They also transmit many disease producing organisms. However, these are not commonly found on water-fowls as in chicken.

2.8 QUAIL FARMING

Quails are very small sized birds. An adult quail weights between 150 to 200 g and an egg weights around 7 to 15 g. Female quails start laying eggs within 6 to 7 weeks of age and continuously lay one egg daily. They lay about 300 eggs in their first year of life. After that they produce about 150 to 175 eggs in second year. Egg production gradually decreases after their first year of laying period.

Benefits of quail farming

1. Quails are smaller sized birds, so they can be raised within small place. 6 to 7 quails can be raised in the same place that is required for one chicken.
2. Feeding cost of quails is comparatively lower than chickens or other poultry birds.
3. Diseases are less in quails, and they are very hardy.
4. Quails grow very fast and gain maturity faster than any other poultry birds.
5. They start laying eggs within their 6 to 7 weeks of age.
6. It takes about 16 to 18 days to hatch their eggs.
7. Meat and eggs of quail are very tasty, delicious and nutritious. Quail eggs contain comparatively more protein, phosphorus, iron, vitamin A, B1 and B2 than other poultry eggs.
8. Quail farming needs small capital, and labour cost is very low.
9. Quail egg contains 2.47 % less fat than chicken egg. Fat is also very low in quail meat. So quail meat is very suitable for blood pressure patients.

10. As the quail eggs are smaller in size, so the price is also lower than other poultry egg. So marketing is not a problem.

2.8.1 Quail Breeds

A number of quail breeds are available, which are very suitable for profitable quail farming. Popular layer quail breeds include Tuxedo, Pharaoh, British Range, English White and Manchurian Golden. Popular broiler quail breeds are Bobwhite and White Breasted. Besides these exotic breeds a few breeds like Uttam, Ujjawal, Sweta and Pearl have been developed by Central Avian Research Institute, Izatnagar which are very popular.

2.8.2 Quail Housing

Housing is very important for quail farming. Quail can be raised in both litter and cage systems. But quail farming in cage system is more suitable than raising them in deep litter system. In cage system, management is very easy and diseases or other problems are less. There should be proper ventilation system and proper flow of air and light inside quail shelter. 50 quails can be raised in a cage measuring 120 cm length, 60 cm wide and 25 cm height. Wire net is used for making the cages. Measurement of the net would be 5 mm × 5 mm for adult quails. Plastic cages are most convenient for quail farming business. The house must have to be out of the reach of wild animals.

2.8.3 Quail Feeding and Management

An adult quail consumes about 20 to 25 g of food daily. Chick feed should contain 27% and adult feed 22-24% of protein. The feed is comprised with broken wheat, sesame oilcake, rice bran, broken oyster shell, salt and mineral mix.

Quails never incubate their eggs. So, chicks are produced by hatching their eggs through chickens or artificially through using incubators. Incubation period for quail egg is about 16 to 18 days. For maximum egg production, 16 hours of lighting period is required daily inside the quail house. Newly born quail chicks are kept in a brooder house. Chicks need artificial heat and temperature management system for 14 to 21 days from their birth. Adequate fresh and clean water should be provided to the birds according to their demand.

Diseases are less in quails, compared to other poultry birds. Generally they are not provided any disease preventive vaccines. Quail chicks get affected by disease, if they experience sudden temperature or weather changes. If the quails get affected by coccidiosis disease, coaxial 20 are served to the birds by mixing with water (2 g/l) for three days. For ulcerative enteritis (caused by infection with *Clostridium colinum*) one gram streptomycin is mixed with one litter water and served to the quails for three days. The cages or shelters of the quails should be kept dry and clean. Different aged quails are kept separately from each other. The disease affected quails are kept separately from the healthy ones.

2.9 GUINEA FOWL FARMING

Guinea fowl farming is a low-cost enterprise and it needs low-maintenance. Thus it has the potential for landless labourers and marginal and small farmers of arid and semi arid region. Three available varieties are Kadmbari, Chitambari and Swetambari. Weight of the fowl at 8 weeks becomes 500-550 g and at 12 weeks 900-1000 g. The female fowl lays egg at 230-250 days age. Average egg weight is 38-40 g. Egg production in one laying cycle from March to September is 100-120 eggs. The correct way to catch a guinea is to clap both hands against its wings. Once caught, the bird is carried by its strong wings, never by its brittle legs.

Benefits of guinea fowl farming

1. Guinea fowl are highly free moving birds and possess excellent foraging capabilities with unique ornamental value.
2. They act as bioinsecticides and bioherbicides. Working as a team, guineas will eat any pest they can get their beaks on, but unlike chickens, do so without tearing and scratching the garden plants. Since the birds are free-range, these will hunt ticks (or beetles, fleas, grasshoppers, crickets, snakes) all around. These are a more natural option to control the insect pest population than pesticides.
3. Guinea fowl are hardy and resistant to many common diseases of other poultry birds.
4. Guinea fowl are used as a watch dog in protecting the farm flock from intruders and predators by its loud, harsh, cry and its pugnacious disposition.
5. Guinea fowl are extremely good runners which help them to escape from predators.
6. Hard egg shell provides minimum breakage and long keeping quality.
7. Birds tolerate weather extremes fairly well after they are fully feathered and can be suitable to any agroclimatic conditions.
8. Input requirements are low. Birds consume all non-conventional feed stuffs, usually not used in chicken feeding. They are more tolerant to mycotoxin and aflatoxin. No requirement of elaborate and expensive housing.

9. Guinea meat is rich in vitamins and low in cholesterol. Guinea meat is tastier than chicken and it contains less fat and fewer calories.
10. The nutrient-rich dropping of the birds can be composted and used in the garden.
11. The feathers of guinea fowl are used in arts and crafts such as hat and head gears.

2.9.1 Guinea Fowl Housing

Domestic guinea fowl are not fully domesticated and are still a bit wild. They are not used to staying inside a house. They should be kept on free range during the day and kept inside a house at night to be protected against predators. Guinea fowl will get sick and die if they are kept in a wet or dirty house. The guinea fowl house of 3 m × 3 m is large enough to keep as many as twelve adult guinea fowls comfortably. The house should have a well maintained grass thatch roof, covered with plastic sheeting to ensure that it does not leak. It should be well ventilated with two opposite windows, each 0.75 m × 0.25 m. The birds need perches. At least one bamboo perch should be faced to the walls of the house at about 1 m from the floor. The floor of both the house and the play area should be covered with a thick layer of sand (15 cm deep). Sand is the best bedding material for guinea as it provides them with a source of grit. Droppings should be swept out once a week. The sand should be changed once a month.

A baby guinea is called a 'keet'. Keets are less tolerant of crowded and dirty conditions and will die also. Young keets are very vulnerable to wet conditions and will certainly die if they get wet. They need to be kept inside the house if it is raining or if the ground outside is wet. The baby guineas become acquainted with their home ground as they broaden their foraging range, while adult guineas are likely to fly the coop the first chance they get. By the time keets are one month old, they develop their first set of feathers.

2.9.2 Guinea Fowl Feeding and Management

Domestic guinea fowl thrive and breed well when they are allowed on free range during the day. They usually move together in groups and scratch for food such as insects, snails, seeds and berries with their feet or their bill. Although they do pick up food while on free range, they should still be fed each day. By providing food and water inside the guinea fowl house, the guinea fowl are trained to return to their house each evening. Ideally, food for guinea fowl should be easily available, relatively cheap and nutritionally balanced. Guinea fowl need to be fed a well-balanced diet in order to gain weight and produce more eggs. A balanced diet consists of a mixture of different types of food which contain protein, fats, carbohydrates, minerals, vitamins and water.

As the keets are not safe to graze in the wild it is important that they should be given plenty of feed while they are inside the brooder. Simply feed is scattered on a newspaper and they will peck and feed for hours. As for feeding the keets, a 22% protein ration consisting of either commercial turkey starter or mashed hard-boiled eggs mixed with cottage cheese and a little cornmeal or oatmeal is given. After four weeks, 18% grower ration, or a mixture of cracked corn and whole wheat supplemented with young grass clippings, chopped lettuce, and other fresh greens are given. Water is always provided to the birds.

Guinea fowl only lay eggs in the breeding season which is from September to April. Most eggs are laid between the months of November and January. Guinea fowl are prolific layers and will lay many eggs if they are encouraged to keep on laying eggs by removing their eggs each day. Each female guinea fowl lays on average about 4 eggs a week or 17 eggs each month. She will continue laying that number each year until she is five years old.

2.10 PIGEON FARMING

Pigeon farming requires less labour and low investment. The meat of baby pigeon (squab) is very tasty and nutritious. Squabs have huge demand. Generally pigeons are raised in pair. One pair of male and female pigeon stays together for their whole life. Male and female both collect straw together and build a small nest for them to live.

Benefits of pigeon farming

1. It is very easy to handle the pigeons.
2. From their six month of age they start laying eggs and produce two baby pigeon per month on an average.
3. Pigeon can be raised easily in the home yard and roof of the house. A pigeon house can be built in a small place with little investment.
4. Baby pigeon become suitable for consumption within their 3 to 4 weeks of age.
5. Pigeon feeding costs is very low. In most cases they collect food by themselves.
6. Pigeon meat is very tasty and nutritious. The squab has a great demand in the market as a patient's diet.
7. Diseases are comparatively less in pigeons.
8. Different types of toys can be made by the feather of pigeons.
9. Pigeons help to keep the environment safe by eating different types of insects.

10. Pigeon farming is suitable for landless, marginal and small farmers and can be integrated to any farming system.

2.10.1 Pigeon Feeding and Management

Pigeons generally eat wheat, maize, paddy, rice, legume, mustard, gram, etc. Foods are kept in front of their house and they will take the food by themselves. Balanced feed is required for proper growth, good health and for high production. Commercial chicken feed can also be fed to the pigeons. Nutrient requirement of pigeon is crude protein (13.5%), carbohydrates (65%), crude fibre (3.5%) and fat (3%). For fast growing of baby pigeon and for nutrition of adult, oyster shell, lime stone, bone powder, salt, greet mixture, mineral mixture, etc. are also fed to the birds with their regular feed. Along with this, some green vegetables are fed them daily. The daily feed intake of an adult pair is about 120 g.

Baby pigeons (squab) do not need extra feed for 5-7 days. They take crop milk from their parents' stomach which is known as pigeon milk. The pigeon milk looks like thick custard and has high protein content. Male and female pigeon feed their baby in this way for 10 days. After that, they become able to fly and feed themselves by their own. Pigeons require water for drinking and for bathing to remove external parasites. A pair of pigeons drink about 200 ml of water/day.

Female pigeon start laying eggs at the age of 5-6 months of age. They lay two eggs every time and their breeding capability stays for about 5 years. Both male and female pigeons hatch the eggs. Usually it takes about 17 to 18 days to hatch the eggs. Female pigeon feed their baby for ten days by their lips. After that, they start taking supplementary food by their own. The young ones become independent in 35 days. Squabs of 30-35 days age are best suited for meat purpose since recovery is high with body weight of 350 g. In addition to meat, pigeon droppings (25 g/bird/day) can be used as fish feed or enriched manure.

Diseases in pigeons are comparatively less than any other poultry birds. They suffer by TB, paratyphoid, cholera, pox, influenza, etc. To keep the birds healthy the pigeon house should be kept clean and germ free. The disease affected birds should be separated from healthy ones.

2.11 TURKEY FARMING

Turkey occupies an important position next to chicken, duck, guinea fowl and quail among the poultry birds in India and forms only about 2% of the total poultry population. Turkey farming is getting popular fast in southern regions. They are reared for meat only and its meat is the leanest among other domestic avian species. Turkey grows faster like broiler chickens and become suitable for slaughter within a very short time.

2.11.1 Turkey Breeds

There are three types of turkey commonly available in India. They are Broad Breasted Bronze, Broad Breasted White and Beltsville Small White. Broad Breasted White is a cross between Board Breasted Bronze and White Holland with white feathers. White turkeys seem to be more suitable for Indian conditions as they have better heat tolerance and also good and clean in appearance after dressing. Virat, a variety of Broad Breasted White type turkey has been developed by Central Avian Research Institute, Izzatnagar. Indigenous and non-descriptive turkeys are found in good numbers in Kerala, Tamil Nadu, eastern districts of Uttar Pradesh and some other parts of India.

2.11.2 Turkey Housing

Housing of turkeys may be of range system, confinement or a combination of both. There is a growing trend towards confinement housing. Good housing with all types of essential facilities available is very necessary for commercial turkey production. A permanent house like chicken house may be made and availability of all necessary facilities should be ensured. Turkeys are large in size and become very strong. So, the fencing should be strong enough to protect the birds. Proper protection for the birds from all types of predators and harmful animals should be ensured. Turkeys require a floor space of 0.4 m^2/bird in confinement and 32.5 m^2/bird in range rearing. Turkeys are not allowed grazing in the pasture until they reach grazing age of eight weeks. There should be sufficient flow of fresh air with proper ventilation system and light inside the house. The house should be cleaned regularly. In case of fencing, it must have to be at least four feet above the ground.

2.11.3 Turkey Feeding and Management

Turkeys have relatively higher feed and nutrient requirement as they have much higher growth rate than other poultry birds. Turkeys can be fed with all types of dry feed comprising pellets, wet mash, grain and green feed in well balanced combination. Turkeys need more protein in their food for first few weeks than other domestic birds. Chick starter which contains about 28% of protein can be fed for first six weeks and after that period chick grower feed which contains 20% protein can be fed. Along with providing nutritious food, sufficient amount of fresh and clean water is provided to the birds.

For natural flocks of 12 hens 1 tom (adult male turkey) is kept for breeding. The birds should be carefully segregated when they reach the mating stage. Mating takes place during summer. The males should be kept away from their mates in the morning and selectively mated with birds by turn. Mating in the evening increases the number of fertile eggs. Generally turkey lays egg on alternate days from six months onwards. In a year one hen lays about 150 eggs.

The bird attains a weight of 5-8 kg in a year. Turkey meat has nutritional and sensorial properties which make it almost ideal raw material for rational and curative nutrition. Turkey meat is preferred because of its leanest nature. The protein, fat, energy values of turkey meat are 24%, 6.6%, 162 calories per 100 g of meat. Minerals like potassium, calcium, magnesium, iron, selenium, zinc and sodium are present. It is also rich in essential amino acids and vitamins like niacin, vitamin B6 and B12. It is rich in unsaturated fatty acids and essential fatty acids and low in cholesterol.

2.12 EMU FARMING

Emu farming in India is emerging as a popular and lucrative business model. These are large sized poultry birds of ratite group and have high economic value. They can adopt themselves with almost all types of agroclimatic conditions. They can be raised in both extensive and semi-intensive farming systems. Emus are very strong and live for long time. Mortality rate and other health problems are less in emus.

Benefits of emu farming

1. Emu meat is very healthy and tasty. It is lower in fat and cholesterol and higher in protein and energy.
2. Emu products like eggs, meat, skin, oil and feathers have a high value in the market.
3. Emus take less food and convert them to various types of valuable products.
4. Marginal and small farmers can easily raise some emus in their household with other livestock animals and poultry birds.
5. Diseases are less in emus and they can survive in almost all types of agroclimatic conditions. Indian climate is also very suitable for commercial emu farming.
6. Emu farming business is very profitable and it can be a great source of income and employment for the unemployed youths.
7. Emu farming does not require high technical and management knowledge and skills.

2.12.1 Care and Management of Emu

A newly born emu chick weights around 370 to 450 g. The chicks are kept inside the incubator for first 2 to 3 days. Then the chicks are kept in deep litter floor house. The litter can be prepared with paddy husk and then the litter is covered with new gunny bags. For the first 3 weeks 0.4 m^2 space is required per chick. Under the brooder, sufficient number of feeder and water pots should be kept. To avoid straying and jumping of chicks a chick guard of at least 75 cm height is provided. A 40 watt bulb is sufficient for covering 10 m^2 area. The brooder area of chicks is extended after their 3 weeks of age by widening the chick guard circle. The chick starter mash is fed for about 14 weeks or until they reach about 10 kg body weight. Emu chick requires sufficient run space for proper growth and healthy life. Generally, 9 m × 12 m floor space is suitable for raising about 40 chicks. The birds need big sized waterers, feeders and large floor space. The birds grower mash is fed to the growing birds for their 34 weeks of age or until they reach 25 kg body weight. Along with grower mash about 10% of greens are added. In grower stage, 12 m × 30 m space for 40 birds is required. The chicks or grower are never kept with adult emus.

Generally, emu birds become mature at their 18 to 24 months of age. For proper breeding purpose one male is paired with one female bird. 240 m^2 floor space is required for every pair during mating. More vitamins and minerals are added in their food and fed to the birds before 3-4 weeks of breeding. Generally an adult bird eats about 1 kg food daily. But during mating period food taking rate get reduced drastically. A female emu lays about 15 eggs during first year and this number gradually increase from the next year. They lay eggs for the first time at one and half years of age. An egg weighs around 475 to 650 g. The eggs are like tough marble and are of greenish coloured. Normally an emu egg takes about 52 days incubation period for producing a chick.

2.13 FISHERY

The demand for fish is always high since fish is one of the favourite items in the food menu of about 60% of Indians. The ever increasing demand for fish can be met through inland freshwater fish farms.

Advantages of fish farm business

1. Market demand and price is always high for fish and fish related products in India.
2. Indian climate is very suitable for fish production.
3. Various types of easily found water sources are available.

4. Various types of fast-growing fish species are available. Farming those fast-growing fish species ensure rapid returns from the investment.
5. Integrated fish farming with animals, birds, and vegetables reduces feeding costs and ensures maximum production.
6. Fish farming is comparatively low risk enterprise.
7. Commercial fish farming can create new income and employment opportunities. Unemployed educated rural youths can start raising fishes. This will provide them both financial freedom and permanent income opportunities.
8. People with other occupation or job can also start fish farming without impacting their regular work. It needs comparatively less attention than that of the livestock farming.
9. Incentives and credit facilities are available to the fish farmers.

2.13.1 Preparing the Pond

The most important infrastructure of the fish farming is a pond. Fish can be raised in both seasonal and permanent pond. In ponds where water does not exist for the whole year, some fast-growing and quick maturing fish breeds can be raised. The optimum size of the pond is rectangular with size varying from 0.1 to 2.0 hectares with a depth ranging from 2.0 - 3.0 m.

The site should contain soft bottom soil or mixed soil comprising of clay sand and silt to ensure good water bearing capacity as well as production of natural food organisms. In case of mixed soils, clay content up to 30% is desirable. Water temperature 25-30° C and dissolved oxygen level 5 mg/l is conducive for fish growth. For carp culture, the water depth should be between 1.5 and 2 m depth. For nursery pond management the water depth should be maintained between 0.6 m and 0.8 m.

Soil and water: Loamy or clay loamy soil is suitable for fish pond. Soil *p*H should be near neutral, i.e. 6.5 to 8.5. Though the soil type cannot be changed except in the long range plans, the *p*H has to be brought to neutral if the pond soil and water are saline, alkaline, sodic or acidic.

Aquatic weeds: The aquatic weeds not only take away the nutrients but also upset the oxygen balance in the water by releasing CO_2 into the pond during the night. Aquatic weeds also obstruct the movement of fishes and the netting operations. The aquatic weeds may be free floating surface weeds, submerged weeds, rooted emergent weeds, marginal shallow water weeds and algae. The fish pond should be free from these weeds.

Unwanted fishes: The unwanted fishes in the ponds may be predatory or weed fishes. They compete with cultured fish for feed, nutrients and space. These predatory and weed fishes can be eliminated through repeated netting of the pond. Otherwise the whole water from the pond is drained out and the unwanted fishes can be eliminated manually and then the pond is refilled with water.

Pre-stocking management: Liming of pond is done @ 250 kg/ha about 15-20 days before stocking of fishes as it is essential to maintain the *p*H of water and also to help maintain the hygienic condition of water. Manuring has to be done after filling the pond with water for production of natural food organisms, i.e. plankton. The amount of fertilisers required in general for fish ponds is 5-10 t/ha/year of cow dung, 200 kg/ha/year of urea, 150 kg/ha/year of single super phosphate and 40 kg/ha/year of muriate of potash. The natural productivity is maintained by the regular manuring and fertiliser application in the pond so that all essential nutrients for the growth of aquatic micro and small organisms (both plant and animal types) are supplied which directly or indirectly serve as feed for the fishes.

2.13.2 Stocking and Post-stocking Management

Composite fish culture: About 15-20 days after the initial manuring selected species of the carps are introduced into the pond. When several species of fishes are reared together in the same pond in an intensive way it is called composite fish culture. Composite fish culture should be a combination of phytophagous fish (catla, rohu, mrigal), plankton feeder (silver carp), omnivorous (common carp) and mud eaters (mrigal and calbasu) in a suitable proportion to maximise the productivity with the available resources in the system. Depending on the number of species the ratio of the species will vary. Catla and silver carp are surface feeders while rohu is a column feeder. Mrigal and calbasu are bottom feeders. Composite fish culture should have all three types of feeders in a ratio of 40:30:30 for surface, column and bottom feeders.

Stocking size and rate: The survival of the fingerlings depends on their size, bigger the size greater will be the survival rate. The fingerlings stocked should have a size of 7.5 to 10 cm. The stocking rate depends on the volume of the water and on the oxygen balance of the pond rather than the size of the pond. The desirable stocking rate is 6000 fishes per hectare.

Time of stocking: From the temperature point of view the best time to stock the pond is when the water in the pond is within the optimum range of 20 to 30° C. Temperatures below 18° C and above 30° C will affect the growth of the fish. Hence stocking is avoided in winter and summer months.

Feeding of carps: Feeds for the carps may be of natural feeds and artificial feeds. The natural growth of flora and fauna in the pond can be increased by regular manuring. However, for commercial fish farming, supplementary feeding should be provided to the carps. The feeding of carps also is easier as they can be fed on most of the by-products like rice polish, wheat bran, oil cakes, vegetable wastes and other farm wastes. Rice bran and oilcake mixture at 1:1 ratio should be prepared and if possible vitamin and mineral mixture can be added.

Water management: Proper depth of the water should be maintained always in the pond. An average depth of 2.0-3.0 m depth should be maintained in the case of six species composite fish culture. Care should be taken to avoid decomposition of large amount of organic matter at the bottom of the pond. Water quality should be periodically checked to ensure that it is within the suitable range especially for *p*H and dissolved oxygen content in water. If required, aerators can be used to maintain desired level of oxygen in the fish pond.

Disease management: Fish diseases can be treated by treating water using salt, potassium permanganate solution, chemicals, etc. It is better to prevent diseases than curing it.

Harvesting: Harvesting can be done either by partially draining water out of the pond through an outlet point, or by repeat netting. Usually the fishes are harvested after 8-12 months of stocking. Shorter rearing periods may also be resorted to depending on the pond conditions and size, stocking density, and demand and price of fish in the local markets. An individual fish grows to the size of 0.8-1 kg in one year.

2.14 APIARY

Beekeeping or apiary has been evolved as an excellent enterprise for rural poor. The direct benefits from apiary are hive products like honey, bee wax, pollen, bee venom, royal jelly whereas indirect benefit is from pollination of different crops. The value of additional yield obtained due to pollination is 15-20 times more than the value of all the hive products put together. Beekeeping is most suited to landless people or with small land holdings. This enterprise does not compete with any branch of agriculture.

2.14.1 Scientific Beekeeping

For sustainable product of honey and other hive products, the following steps should be followed in scientific beekeeping.

Training: Scientific beekeeping helps to protect the bees and simultaneously harvest honey of good quality. Beekeeping is very easy but it involves lot of skill, requires clean understanding of bee behaviour and their language. So training is essential for beekeeper.

Knowledge about honeybees: The grower should get acquainted with various bee species like Indian hive bees, Italian bees and stingless bees and the queen, drone and worker of their colonies. Knowledge must be acquired about the working style of the different honeybee species and their division of labour. Only two species have been truly domesticated *Apis mellifera* (European honeybee) and *Apis cerana indica* (Indian honeybee).

Selection of site: Neat and clean area with plenty of fresh water, well ventilation, ample shade and sunshine, away from main road but well communicated and least disturbed area is very ideal for beekeeping. Site selected for apiary should have sufficient bee flora within a radius of 1 km and provide nectar and pollen to the bees round the year. Site should be away from water logged areas, cattle shade, kitchen/smoky area.

Procurement of bee equipments: ISI 'A' type beehives with 8 frames for Indian hive bees, langstroth hive with 10 frames for Italian honeybees and wooden box of 25 cm × 15 cm × 13 cm size for hiving the stingless bees are necessary. Smoker, honey extractor, queen gate, bee veil, gloves, queen excluder sheet, drone trap, knife, and hive tool should be procured prior to beekeeping.

Procurement of colony: To start the beekeeping a colony may be procured by dividing an old colony or capturing a swarm in the season. Beginners may also purchase the bee colonies from a progressive beekeeper. The colony should be of four frame colony with young (< 1 year) queen, free of drones and sufficient workers. The frame should have sufficient eggs, broods, good nectar and pollen store. Beekeeping should be initiated in the beginning of honey flow season. The ideal time either for procurement or purchase of colony is December-May.

Seasonal management: Regular inspection of the hive at 10 to 15 days interval round the year is required. Provisioning of heat in winter, protection from rainwater in monsoon and shade in summer are prime steps in seasonal management. Dividing strong colonies, swarm control, mounting super chamber over brood chamber, regular honey extraction, forcing bees for construction of new comb, uniting the weaker colonies is necessary during spring. Providing artificial diet, prevention of robbing and management of laying workers during dearth and queenless colony should be done immediately.

Protection from enemies and bee disease: Knowing the enemies of bees like wax moth, wasps, mites, ants, spiders, viral diseases, macroorganisms and their symptoms, appropriate preventative/curative method should be adopted. Maintenance of strong colony and keeping bottom board clean reduces occurrence of enemies and diseases in bees.

Honey extraction, processing and storage: Honey is extracted when 70-80% of comb in super chamber is filled with honey and sealed. Honey should not be kept exposed for a longer period during harvesting.

2.15 MUSHROOM CULTIVATION

Mushroom is a fruiting body of fungi. Mushrooms need substrate to grow, a mushroom substrate is a substance where mushroom mycelium will grow. Mushroom cultivation is landless farming using various substrates like paddy straw, wheat straw, sawdust, and other agricultural residues that are rich in carbon and nitrogen sources. Mushrooms require less growing materials, water, and energy than other types of crops. Mushroom farming emits much less carbon dioxide. Mushrooms require less land than other agricultural practices because they are stacked vertically in growing facilities.

Mushrooms are a good source of nutrition like protein, carbohydrate, minerals, vitamins, fat, ash and fibre. Mushrooms also contain a high amount of calcium, phosphorus, sodium, and potassium and low but available form of iron and magnesium which shows good health effects. Mushroom contains vitamins like thiamine, riboflavin, niacin, biotin, and vitamin C. Mushrooms have been utilised for food and medicine as they contain different primary and secondary metabolites, showing beneficial health effects. The demands of mushroom in pharmaceutical industry increase due to its nutraceutical application. Mushrooms are being used for the production of various value-added products such as mushroom *papad*, mushroom pickle, mushroom biscuits, mushroom beverage, etc.

2.15.1 Mushroom Production Technology

Paddy straw mushroom (*Volvariella volvacea*), oyster mushroom (*Pleurotus ostreatus*) and button mushroom (*Agaricus bisporus*) are the three major types cultivated in India. Paddy straw mushrooms can grow in temperatures ranging from 35 to 40° C. Oyster mushrooms on the other hand, are grown in the northern plains while button mushrooms grow during the winter season. All these mushrooms of commercial importance are grown by different methods and techniques. Mushrooms are grown in special beds known as compost beds.

2.15.1.1 Paddy straw mushroom

Straw bundles of 60-90 cm length are soaked in clean and cold water for 6 hours. Period of soaking depends upon the stiffness of the straw. Then substrate (soaked straw) is pasteurised physically or chemically for 1 hour. In physical method it is treated in hot water or steam pasteurised at 70-80° C for one hour.

Alternatively, the bundles are soaked in water containing 1-2% $CaCO_3$ powder for the required period so that the *p*H of the medium is improved. This suppresses the growth and multiplication of moulds in the substrate. Then bundles are kept in a slanting manner upside down to drain out excess water.

Four bricks are placed at 60 cm apart from each other and bamboo sticks are put on it to make a platform. The spawn is removed from the spawn bottle by breaking it and divided into 4 parts. For preparing a bed of 45 cm × 45 cm × 45 cm size, 200 g of spawn, 200 g organic supplement such as chokada or pulse powder and 7 kg straw are required. First layer of straw having about 12-13 cm thickness is spread over the platform. One fourth of the spawn is put at 8 cm apart from the periphery at spacing of 8 cm also. One-fourth of the pulse powder should be sprinkled on the spawn bits. After the 1st layer is complete, another layer of straw 12-13 cm thickness is laid opposite to the 1st layer and spawn along with organic supplement are sprinkled. Then the 3rd layer is prepared just like 1st layer and seeded 2 parts of spawn as well as organic supplements both at edges and centre. A thin layer of 5-8 cm straw is spread on the 3rd layer, which is the cover layer. In this way the bed is prepared and polythene of 1.8 m × 1.8 m is covered in order to maintain moisture for 7 days which is known as incubation period. Covering of polythene helps in rising of temperature inside the bed, less accumulation of oxygen and contaminant moulds with the bed. After completion of incubation period the bed is covered with fungal mycelium.

On 9th and 10th day, pinheads come out from the bed which develop gradually and attain harvestable stage (egg stage) towards 14th and 15th day of spawning. The first flush yields about 80% of the total sporophores and rest 20% is obtained after one week of 1st harvest.

2.15.1.2 Oyster mushroom

Paddy straw is cut into small pieces of 3-5cm length, soaked in clean for 6 hours water and dried in shade till it is neither too dry nor too wet. Two holes of one cm are made in the centre of a polythene bag of 60 cm × 30 cm size (both side open). The bottom of bag is tied with a thread to make a flat bottom. The bag is filled with straw to 10 cm height and then spawn is inoculated. 4-5 layers of straw and spawn are prepared alternatively. Last layer ends up in straw of 10 cm height. The bag is kept in a spawn running room maintained at a temperature of 22-28° C with relative humidity 85-90%. Under favourable environmental conditions within 15-20 days of spawning, the bed surface is covered with the cottony growth of the white mycelium. This condition is called spawn run. When spawn running is completed, the polythene bag is cut opened and taken to cropping room.

Mushroom pinheads appear on 3rd day of opening of beds and mature in 3 days. Matured mushrooms are harvested daily or alternate days, before spraying water. Second and third harvest can be obtained after scraping the surface of beds after first or second harvest. Mushroom yield is 0.5-1.0 kg per bag.

2.15.1.3 Button mushroom

It requires a complex method of preparing compost, which is used as a substrate for mushroom production. Spawning can be done by three methods such as surface spawning, layer spawning and trough spawning. The trays are filled with compost and then spawning is done. After spawning, compost is pressed hard to make it compact. Trays are arranged in the cropping room in tiers and covered with newspaper sheets spayed with 2% formalin. Temperature of 20-25° C and relative humidity of 90-95% are maintained.

After spawn running is completed in 15-20 days, casing is done. Casing is a top-dressing applied to the spawn-run compost on which the mushrooms eventually form. Clay-loam field soil, a mixture of peat moss with ground limestone, or reclaimed weathered, spent compost can be used as casing. Pinheads appear within 10-15 days after casing. Cropping continues for 60-75 days. Mushrooms can be harvested at button stage. Mushroom yield is 6-7 kg/m^2.

2.16 SERICULTURE

Sericulture or silk farming is the cultivation of silk through rearing of silkworm. It involves the raising of food plants for silkworm, rearing of silkworm for production of cocoons, reeling and spinning of cocoon for production of yarn for value added benefits such as processing and weaving.

The production of silk generally involves two processes; care of the silkworm from the egg stage through completion of the cocoon, and production of mulberry trees that provide leaves upon which the worms feed. Cultivation of mulberry plants is called moriculture. Out of 20 species of mulberry four (*Moris alba*, *Moris indica*, *Moris serrata* and *Moris latifolia*) are commonly cultivated for production of leaves. A well maintained mulberry crop yields 30-40 t leaves/ha/year. The crop can yield for about 12-15 years.

India has the unique distinction of being the only country producing all the five known commercial silks; mulberry silkworm (*Bombyx mori*), tropical tasar silkworm (*Antheraea mylitta*), oak tasar (*Antheraea pernyi*), eri silkworm (*Philosamia ricini*) and muga silkworm (*Antheraea assami*), of which muga with its golden yellow glitter is unique and prerogative of India. Mulberry sericulture is mainly practised in states such as Karnataka, Andhra Pradesh, Assam, West Bengal, Jharkhand and Tamil Nadu who are the major silk producing states in the country. India is the second largest producer of silk in the world, next to China.

Each silkworm will eat mulberry leaves until it is about 10,000 times heavier than when they were hatched. They will grow and molt four times before they are ready for the next phase of their lifecycle. It is at this point that these caterpillars create their cocoon by secreting a dense fluid composed of proteins. This fluid is secreted in one long, continuous strand that solidifies upon contact with air to create a single strand of silk that is wrapped around the cocoon thousands of times. A single strand can be more than one km long and only 10 micrometers in diameter. Then this silk strand must be carefully harvested, a process that requires careful timing. If the silkworm undergoes metamorphosis and matures into a silkmoth, it will escape the cocoon by secreting an acidic fluid which dissolves a hole in the cocoon, splitting the silken fibre into short fragments that cannot be reeled or used for silk yarn. Before this happens, the silk must be harvested. To harvest the silk, the cocoon is immersed in boiling water. This process kills the silkworm pupae, but also frees the silk filaments from the tightly wound cocoon and readies them for reeling. Each strand is combined with strands from other cocoons to create a single thread of silk that can be used to create textiles. One thread contains up to 48 silk filaments which is then wound onto a reel, ready to be dyed and eventually used in the fabrics and threads used to create silk embroideries.

The cocoons required for furthers rearing are kept separately and moths are allowed to emerge from them.

2.17 VERMICOMPOSTING

Vermicomposting is a method of preparing enriched compost with the use of earthworms. It is one of the easiest methods to recycle agricultural organic wastes and to produce quality compost. Earthworms consume biomass and excrete it in digested form called worm casts. Worm casts are vermicompost. The casts are rich in nutrients, growth promoting substances, beneficial soil micro-flora and having properties of inhibiting pathogenic microbes. Vermicompost is free flowing, easy to apply, handle and store and does not have bad odour. It is free from pathogens, toxic elements, weed seeds, etc. It is also economically viable and environmentally safe nutrient supplement for organic food production.

2.17.1 Earthworm Species

The surface dwelling earthworms are used for vermicomposting. The African earthworm (*Eudrilus eugeniae*), red earthworm (*Eisenia foetida)* and composting worm (*Peronyx excavatus*) are promising worms used for vermicompost production. All the three types of worms can be mixed together for vermicomposting.

These species are prolific feeders and can feed upon a wide variety of degradable organic wastes. They exhibit high growth rate. The African worm is preferred over other two types, because its multiplication is high and thus produces more vermicompost in short period of time. But red earthworm has a wider tolerance for temperature than the other two species which allows the species to be cultivated in areas with high temperature (often as high as 43° C as well as those with lower soil temperature (often below 5° C).

Factors influencing culturing of earthworms

Several factors control the culturing and maintenance of healthy earthworm populations, of which the most important are food, moisture, temperature, light, pH and protection from predators.

Food: Fresh green matter is not easily fed upon by the worms. Decomposition by microbial activity is essential before earthworms can feed on fresh waste. Earthworms find it difficult to survive when the organic carbon content in the waste is low. So high carbon: nitrogen ratio is required.

Moisture: Moisture levels have to be maintained at around 60% so that the microbial activity is high and the food matter is easy to feed upon. Excess water leads to anaerobic conditions, which in turn lowers the *p*H and creates acidic conditions. Acidic conditions reduce productivity and cause migration of worms.

Temperature: Temperature affects metabolism, growth and reproduction. Soils exposed to the sun lose moisture quickly and are usually devoid of earthworms. Earthworms maintain lower body temperatures than the surrounding soil or organic matter by their metabolic adjustments.

Light: Earthworms are very sensitive to light. The photoreceptor cells detect light and the earthworms move away to avoid strong light.

pH: Earthworms are sensitive to changes in pH. They prefer conditions of neutral reaction. Earthworms find it difficult to survive if the pH falls below 6 and thus they migrate or are killed.

Predators: Earthworms are preyed upon by many species of ants, birds, toads, snakes, moles, cats, rats, dogs, etc. A variety of invertebrates like flatworms and centipedes also feed on earthworms.

2.17.2 Methods of Vermicomposting

Vermicompost can be produced in any place with shade, high humidity and cool. Abandoned cattle shed or poultry shed or unused buildings can be used. If it is to be produced in open area, shady place is selected. A thatched roof may be provided to protect the process from direct sunlight and rain. The waste heaped for vermicompost production should be covered with moist gunny bags.

Vermicomposting can be done either in pits or concrete tanks or well rings or in wooden or plastic crates appropriate to a given situation. Usually, composting is done in the cemented pits of size 2 m × 1 m × 0.75 m. The unit is covered with thatch grass or any other locally available materials. For large scale production pit method of vermicomposting is preferred with pit size of 8 m × 2.4 m × 0.75 m.

Bed method is also common. Composting is done on the floor by making bed (2 m × 1 m × 0.75 m size) of organic mixture. This method is easy to maintain and to practise.

2.17.3 Process of Vermicomposting

Cattle dung, farm wastes, crop residues, vegetable market waste, flower market waste, agroindustrial waste, fruit market waste and all other biodegradable wastes are suitable for vermicompost production. Excreta of pig, poultry and goat are not suitable. Cow dung and chopped dried materials are mixed and are kept for partial decomposition for 30-45 days. In the organic waste mixture cow dung should be at least 30%.

A layer of 10 cm of chopped dried leaves/grasses should be kept as bedding material at the bottom of the bed or pit. The mixed waste is placed into the container up to brim. Each bed or pit should contain 200-250 kg of raw material and the number of beds can be increased as per raw material availability and requirement. Red earthworm 1 kg (about 1000-1500 in number) should be released on the upper layer of bed. There is no necessity that earthworm should be put inside the waste. Earthworm will move inside on its own. Water should be sprinkled immediately after the release of worms. Beds should be kept moist by sprinkling of water daily and by covering with gunny bags. The moisture level should be maintained at 60% throughout the period. Watering should be stopped before the harvest of vermicompost.

Harvesting vermicompost

Compost gets ready in 45-50 days. The finished product is 3/4th of the raw materials used. When raw material is completely decomposed it appears black and granular. In the pit or tank method of composting, the castings formed on the top layer are collected periodically. The collection may be carried out once in a week. The harvesting of casting should be limited up to earthworm presence on top layer. This periodical harvesting is necessary for free flow and retaining the compost quality. Otherwise the finished compost gets compacted when watering is done. In small bed type of vermicomposting method, periodical harvesting is not required. After 2-3 days of harvest, the compost is sieved and packed.

Harvesting earthworm

After the vermicompost production, the earthworms present in the small beds may be harvested by trapping method. Before harvesting the compost, small fresh cow dung ball is made and inserted inside the bed in five or six places. After 24 hours, the cow dung ball is removed. All the worms will be adhered into the ball. Putting the cow dung ball in a bucket of water will separate this adhered worm. The collected worms will be used for next batch of composting.

Nutrient content of vermicompost

The level of nutrients in compost depends upon the source of the raw material and the species of earthworm. Nutrients in vermicompost are in readily available form and are released within a month of application. The available N, P, K, Ca and Mg contents in vermicompost are 0.50, 0.30, 0.24, 0.17 and 0.06%, respectively.

2.18 BIOGAS

Biogas comprises of hydro-carbon which is combustible and can produce heat and energy when burnt. It is a clean, pollution free and cheap source of energy, which is produced by a biochemical process in which certain types of bacteria convert the biological wastes into useful biogas with little investment. A biogas unit is an asset to a farming family. It also produces good manure and improves sanitation. It can be used for cooking purpose, burning lamps, running pumps, etc.

The process of biogas production is anaerobic in nature and takes place in two stages. The two stages have been termed as acid formation stage and methane formation stage. In the acid formation stage, the biodegradable complex organic compounds present in the waste materials are acted upon by a group of acid forming bacteria present in the dung. Since the organic acids are the main products in this stage, it is known as acid forming stage. In the second stage, groups of methanogenic bacteria act upon the organic acids to produce methane gas. Cattle dung is the chief raw material for biogas plants.

2.18.1 Components of Biogas Plant

1. Mixing tank: The feed material (dung) is collected in the mixing tank. Sufficient water is added and the material is thoroughly mixed till homogeneous slurry is formed.
2. Inlet pipe: The substrate is discharged into the digester through the inlet pipe/tank.
3. Digester: The slurry is fermented inside the digester and biogas is produced through bacterial action.
4. Gas holder or gas storage dome: The biogas gets collected in the gas holder, which holds the gas until the time of consumption.
5. Outlet pipe: The digested slurry is discharged into the outlet tank either through the outlet pipe or the opening provided in the digester.
6. Gas pipeline: The gas pipeline carries the gas to the point of utilisation, such as a stove or lamp.

2.18.2 Site Selection

While selecting a site for a biogas plant, following aspects should be considered.

1. The land should be levelled and at a higher elevation than the surroundings to avoid water stagnation.
2. Soil should not be too loose and should have a bearing strength of 2 kg/cm^2.
3. It should be nearer to the intended place of gas use.
4. It should also be nearer to the cattle shed for easy handling of raw materials.
5. The water table should not be very high.
6. Adequate supply of water should be there at the plant site. The plant should get clear sunshine.
7. The plant site should be well ventilated.
8. A minimum distance of 1.5 m should be kept between the plant and any wall or foundation.
9. It should be away from any tree to prevent root interference.
10. It should be at least 15 m away from any well used for drinking water purpose.

2.18.3 Types of Model

The two main designs of biogas plants are the floating gas holder and fixed-dome types.

Float dome type: Khadi and Village Industries Commission have designed two float dome type models such as KVIC vertical and KVIC horizontal. Other popular models include Pragati model and Ganesh model.

Fixed dome type: The gas plant is dome shaped with underground construction. The masonary gasholder is an integral part of the digester called dome. The gas produced in the digester is collected in dome at vertical pressure by displacement of slurry in inlet and outlet. The entire construction is made of bricks and cement. The models available in this category are Janata and Deen-Bandhu.

The size of the biogas plant is decided by the number of family members and the availability of dung. One cubic meter capacity plant will need two to three animals and 25 kg of dung per day. The gas produced will meet the requirement of a family of 4-6 members. It would suffice to have a 2 cubic metre plant to cater to the needs of a family of 6-10 members.

2.18.4 Biogas Slurry

Slurry is obtained after the production of biogas. It is enriched manure with 1.4, 1.1 and 1.0% N, P and K, respectively. Another positive aspect of this manure is that even after weeks of exposure to the atmosphere, the slurry does not attract fleas and worms.

3

Cropping System and Their Classification

In agriculture, the ideal management practices were developed for individual crops through regular research and feedback received from the farmers. However, farmers are growing various crops in different seasons based on their adaptability to a particular season. Thus, selection of crops for a particular situation or region largely depends on climate of that region and soil of the land. Besides, the farmers have their own choices and requirements, and before selecting a crop for cultivation they usually analyse its market demand, economic viability, availability of resources, etc. Very often there is an established and well defined relationship among the crops cultivated by a farmer, either in some form of spatial mixture or in some temporal sequence. Therefore, at present emphasis is being given to develop system based production technology for the crops grown in a year or beyond. This will give a scope to study the relationship among the crops, their environment and the production inputs required.

3.1 CROPPING PATTERN AND CROPPING SYSTEM

'Cropping pattern' and 'cropping system' are two terms sometimes used interchangeably; however, these are two different concepts. While cropping pattern refers to the yearly sequence and spatial arrangement of crops or of crops and fallow in a particular land area; cropping system refers to cropping pattern as well as its interaction with resources, technology, environment, etc. Thus, cropping system is an important component of a farming system and it represents cropping patterns used on a farm and their interaction with farm resources and farm enterprises and available technology which determine their makeup. It is executed in the field level. The term cropping system essentially represents a philosophy of maximum crop production per unit area of land within a calendar year or relevant time unit with minimum natural resource degradation (Gangwar et al., 2012). A cropping system usually refers to a combination of crops in space and time. Combinations in space occur when crops are interplanted and combinations in time occur when crops occupy different growing periods.

Cropping system is a critical aspect in developing an effective ecological farming system to manage and organise crops so that they best utilise the available resources such as soil, air, sunlight, water, labour, equipments, etc. Cropping systems approach enables to address issues pertaining to maximising system productivity on annual basis. It also ensures utilisation of resources with higher efficiency by considering various interactions and direct, residual and cumulative effect in soil-plant-environment system.

3.1.1 Scope of Cropping System

1. Cropping system of a region optimises the use of locally available resources by combining the different components of the farm system, so that they complement each other and have the greatest possible synergetic effects.
2. Cropping system relies mainly on resources within the agroecosystem by replacing external inputs with nutrient cycling, better conservation, and an expanded use of local resources.
3. A suitable cropping system would reduce the use of off-farm, external and non-renewable inputs with the greatest potential to damage the environment or harm the health of farmers and consumers, and a more targeted use of the remaining inputs used with a view to minimising variable costs.
4. Cropping system improves the match between cropping patterns and the productive potential and environmental constraints of climate and landscape to ensure long-term sustainability of current production levels.
5. A good cropping system works to value and conserve biological diversity, both in the wild and in domesticated landscapes, and making optimal use of the biological and genetic potential of plant and animal species.
6. Cropping system takes full advantage of local knowledge and practices, including innovative approaches not yet fully understood by scientists although widely adopted by farmers.

3.1.2 Objective of Cropping System

1. To sustain agricultural production by having healthy rural communities, conserving natural resources, and remaining economically competitive.
2. To identify methods of reducing off-farm input expenses, particularly irrigation, herbicides, and nitrogen fertiliser.
3. To develop cropping systems that implement integrated pest management practices for prolonging the useful life of selective herbicides, and reducing the amount and expense of pesticide applications.
4. To design cropping systems that maximise precipitation use efficiency and fallow use efficiency.

3.2 CLASSIFICATION OF CROPPING SYSTEMS

Depending on the resources and technology available, different types of cropping systems are adopted in an area.

3.2.1. Monocropping

Monocropping refers to growing only one crop on a particular land year after year. Many farmers plant only one crop in the same place year after year due to climatic factors or resource constraints. In low rainfall areas groundnut or cotton or millets are grown in uplands year after year. Similarly, rice is grown, as it is not possible to grow any other crops, in canal irrigated areas, and under water logged conditions.

Advantages of monocropping

1. Monocropping increases farming efficiency of a farmer. When the farmer grows only one type of crop he becomes an expert in that crop and it is a more profitable way to farm than switching crops around each year.
2. Since the crop is the same over the whole farm the crop management decisions become simplified.
3. Need of farm equipments is less. Only the machinery and tools necessary to maintain the crop needs to be purchased.
4. The seasonal work is predictable allowing scheduling of casual labours and machinery.
5. Monocropping reduces the overall cost of production.
6. The production technology becomes simple and much skill may not be required.

Limitations of monocropping

1. If the market falls then income is severely affected due to great dependency on one crop.
2. Monocropping is more susceptible to pests and diseases. One pest or disease can wipe out an entire crop and there is no income for that season or year. To save the affected crop more chemicals are to be used.
3. Different plants have different nutritional requirements. Using the same crop repeatedly year after year will tend to pull soil nutrients unevenly. This will increase the application of higher doses of chemical fertilisers, which may pollute waterbodies by eutrophication.
4. Since more chemicals are used in monocropping systems the toxic residues of pesticides make their way into groundwater or air, creating environmental pollution.

5. Monocropping is usually associated with high levels of mechanisation which leaves the soil compact and the soil structure is distorted. In the long-run soil health is deteriorated.
6. In monocropping the land remains bare between harvest of a crop and planting of the next crop for a substantial period which increases the erosion.

3.2.2 Multiple Cropping

Growing two or more crops on the same piece of land in one calendar year is known as multiple cropping. It represents a philosophy of maximum crop production per unit area of land within an agricultural year. Thus, it is the intensification of cropping in time and space dimensions, i.e. more number of crops within a year and more number of crops on the same piece of land at any given period. Multiple cropping makes effective use of inputs such as soil, water, fertiliser, etc. Thus, output per unit area increases. It can be practised in annual food crops, fodders, vegetables, fruit plants and perennial crops. It includes intercropping, mixed cropping, sequence cropping, crop rotation, relay cropping and paira cropping, etc. For successful multiple cropping, availability of irrigation facility and other input resources including labour should be ensured at right time.

Principles of multiple cropping

1. The crops should be agro-climatically suitable and technically feasible.
2. The most profitable crops capable to utilise growth resources and land in best manner should be chosen.
3. The crop combinations and input rations should be compatible with farmer's skill, enterprise preference, capital and assured irrigation facility.
4. The essential inputs like improved seeds, fertilisers, pesticides and labour should be available locally.
5. Marketing, storage and transportation facilities of the produce are to be ensured before choosing the crops.

Advantages of multiple cropping

1. Land is better utilised through multiple cropping.
2. In multiple cropping, incidence of crop failure owing to biotic agents is minimised. The loss of any crop due to biotic or abiotic stress may be compensated to some extent by another crop.
3. Overall production and yield per unit of area is increased.
4. Cost of production is reduced in multiple cropping as compared to individual crops grown for the same level of production.

5. Pest and disease infestation is checked.
6. Different type of products can be produced at a time in multiple cropping.
7. Multiple cropping provides a balance diet for a farm family
8. Multiple cropping helps in maintaining soil fertility provided suitable crops such as legumes are included in the cropping system. It improves or maintains the soil health.
9. Weed menace is reduced in multiple cropping. It narrows the space available for weeds to grow and sometimes may hamper their growth through exudation of allelochemicals.
10. An important aspect of multiple cropping is the utilisation of nutrients more efficiently as the crops growing on the same piece of land simultaneously would have different nutritional requirements.
11. Multiple cropping is also important from marketing point of view. As more than one product is available simultaneously if the selling price of one product is less in the market, the other will be there to compensate.
12. It generates more employment opportunity by having crops round the year.

Limitations of multiple cropping

1. Multiple cropping may reduce the farming efficiency of a farmer. When a farmer grows different type of crops, he cannot specialise in production technology of all the crops. He has to acquire the latest technologies for the crops grown which may not be possible at all the time.
2. Sometimes the crop management decisions become difficult and often the advice of the professionals may be required.
3. The survival of pests becomes easier in a multiple cropping. Pests can easily shift from one crop to another crop. Sometimes the insect pests and diseases get more favourable environment to flourish, thus diminishing and deteriorating crop yields.
4. Intercultural practices are sometimes difficult to be carried out in the crops grown together.
5. Weed management may also be difficult.
6. Initial investment may be high with procurement of different types of inputs including equipments and machineries for a number of crops at a time.

3.2.2.1 Intercropping

Intercropping refers to growing of two or more dissimilar crops simultaneously on the same piece of land. Practice of intercropping is as old as agriculture. But scientific approach to it is relatively new. The recommended optimum plant population of the base crop is suitably combined with appropriate additional plant density of the associate crop. Various workers have classified and defined intercropping in many ways. However, the objective remains the same, i.e. the intensification of cropping both in time and space dimensions and to raise productivity per unit area by increasing the pressure of plant population. It has better utilisation of growth resources than sole cropping. Intercropping results in more efficient use of solar energy and harnessing benefits of positives interactions in crop associations. These advantages are, in general more pronounced in wide-spread crops and stress environments (Gangwar et al., 2012). Intercropping can be practised as annual plants with annual plants intercrop, annual plants with perennial plants intercrop and perennial plants with perennial plants intercrop.

Types of intercropping

Intercropping may be of various types based on the method of sowing and objective of cropping.

Mixed intercropping: It is an intercropping where two or more crops are grown simultaneously intermingled in the same piece of land with no distinct row arrangement. This type of intercropping can be suitable for grass-legume intercropping in pastures. The mixed intercropping is commonly observed to fulfil the requirement of food and forage where the land resource is a limiting factor. Sometimes it is also referred to as mixed cropping.

Row intercropping: It is an intercropping where at least one crop is planted in regular rows, and crop or other crops may be grown simultaneously in row or randomly with the first crop. The row intercropping is a usual practice targeting maximum and judicious use of resources and optimisation of productivity.

Mixed row intercropping: It is growing two or more crops simultaneously in the same piece of land intermingled within a distinct row arrangement.

Strip intercropping: It is growing two or more crops simultaneously in different strips on sloppy lands wide enough to permit independent cultivation, but narrow enough for the crop to interact with each other. Strip intercropping enhances greater radiation use efficiency in marginal and poor lands. A combination of soil conserving and depleting crops are taken in alternate strips running perpendicular to the slope of the land or the direction of prevailing winds. An important objective of strip cropping is the reduction of soil erosion and harvesting of yield output from poor lands.

Relay intercropping: Growing two or more crops simultaneously during at least a part of the life cycle of each. A second crop is planted after the first crop has reached its reproductive stage but before it is ready for harvest. This practice is beneficial in terms of resolving time conflict and limitation of soil moisture for plantation of various crops. For example delayed sowing is an important reason for yield decline in field pea or chickpea that can be avoided through relay cropping of these crops in standing rice crop. Relay cropping can also fetch certain other benefits such as usage of residual nutrients and moisture from the previous crop and reduced planting costs. This practice is also known as *paira* cropping.

Parallel intercropping: In this practice two crops are selected which have different growth habits and have a zero competition between each other and both of them express their full yield potential. Greengram or blackgram with maize, and greengram or soybean with cotton are the examples of parallel intercropping.

Companion cropping: In companion cropping the yield of one crop is not affected by other. In other words, the yield of both the crops is equal to their pure crops. The standard plant population of both crops is maintained. Wheat, *toria*, potato, etc. can be taken as companion crops with sugarcane.

Additive intercropping: Here the plant population of the main or base crop remains unchanged as a pure crop stand but one or more minor crops are grown simultaneously with it by utilising the inter-row space. Wheat, rapeseed or potato can be taken as intercrops by utilising the inter-space of two rows of sugarcane planted with its normal spacing.

Substitutive intercropping: In the substitutive or replacement series of intercropping, the crops grown together are known as component crops or intercrops. Here, one component crop is introduced by the replacement of the other crop and no crop is sown with its fullest population as seeded in respective sole cropping. In this system, a definite proportion of a crop is sacrificed and the component crop is introduced in that place. Sometimes to obtain yield advantage, plant population may be increased compared to their density adopted in the pure stands. In such an intercropping system, competition among species is relatively less than additive series.

Alley intercropping/alley cropping: Food crops are grown in alleys formed by hedge or shrubs or trees. Gliricidia + groundnut or any suitable annual crop is an example of alley cropping. It is an agroforestry system.

Trap crops: These crops are grown in the main cropped field in definite rows to trap insect pests.

Filler cropping: Growing of short duration crops in between the newly established perennial crops for few years to fill the space and to utilise the resources. Coconut + pineapple, mango + greengram, guava + turmeric are the examples of filler cropping. It is also an agroforestry system.

Principles of intercropping

1. Before putting any intercrop with the main crop it is very essential to know the prerequisites of the companion crops such as soils and water requirement compatibility, competition for space, sunshine and air, compatibility for pests and diseases, duration and yielding potential and time of sowing and harvesting. The selected intercrop should be compatible with the main crop in its water, nutrients and soil requirement.
2. The crops grown in association should have complementary effects rather competitive effects. The subsidiary crop should be of shorter duration and of faster growing habits to utilise the early slow growing period of main crop and they must be harvested when main crop starts growing.
3. The proper spatial arrangement is required to maximise cooperation and minimise the competition between or among the crops selected for intercropping. An intercropping is successful only when a suitable companion crop is selected to grow with the main crop.
4. Overcrowding should always be avoided by optimising the population of intercrops. A standard plant population of main crop should be maintained and plant population of subsidiary crops can be increased or decreased as per demand of the situation.
5. The crops of different maturity dates should be preferred to reduce the competition for resources during peak demand period.
6. In an intercropping system the component crops should have different root depths so that they can draw water and nutrients from different layers of soil. This will minimise competition for nutrients, water and root respiration.
7. The intercrop should not be an alternate host for pest and diseases of the main crop. Component crops of similar pests and disease pathogens and parasite infestations should not be chosen. Crops selected should be of different families to avoid pests and diseases.
8. Component crops should have similar agronomic practices.
9. Erect growing crops should be intercropped with cover crops like pulses. These check the weeds and reduce soil erosion. The losses of water due to evaporation are also reduced.
10. The planting method and management should be simple, less time consuming, economical and profitable so that it may have wider adaptability.
11. Fast growing crops should be taken with slow growing crops.
12. Crops should have least allelopathic effect.

Advantages of intercropping

1. Additional income is realised from the companion crop or crops. Thus, intercropping gives higher income per unit area than sole cropping.
2. If the base crop is damaged due to any unfavourable climatic conditions or spread of insect pests and diseases, the companion crop may give some income. So it acts as an insurance against failure of crop in abnormal year.
3. Legumes grown as companion crops always benefit the principal crop through N-fixation. Legumes also leave a huge quantity of nitrogen rich organic matter in soil in the form of roots and leaf litter.
4. Quick growing companion crops always suppress the growth of weeds thriving in the inter-spaces of the main crop.
5. Intercropping practices encourage better utilisation of growth resources like nutrient, water, light and space.
6. Incidence of insect pests and diseases is low.
7. Employment opportunity is created throughout the year. It offers best utilisation of labour, machine and power throughout year.
8. Soil erosion due to runoff is checked as the soil is covered with vegetation most part of the year.
9. Intercrops maintain soil fertility as the nutrient uptake is made from both layers.

Limitations of intercropping

1. The fertiliser management is difficult in intercropping practice because the nutrient requirement of the crops is different.
2. Harvesting of different crops become difficult as the seeding and harvesting time of the crops vary.
3. If the selection of crops in an intercropping system is not compatible, the growth of one crop may be suppressed by another crop. There may be competitive effects among component crops.
4. Incidence of insect pests and diseases is increased because sometimes one crop may serve as alternate host for pests and diseases.
5. The grower has to acquire the latest technologies for the crops grown which may not be possible at all the time.
6. Inter-cultural practices are sometimes difficult in an intercropping system since the time and requirements of agronomic practices for different crops in a system are different. Mechanisation is also difficult.

7. Application of irrigation water, nutrients, pesticides or herbicides is difficult in an intercropping system and for proper management of the system some skill is required.
8. It is labour intensive.
9. There may be allelopathic effect.

3.2.2.2 Mixed cropping

Growing of two or more crops simultaneously on the same piece of land, without any definite row arrangement is termed as mixed cropping. It is mainly practised in areas where climatic hazards such as flood, drought, frost, etc. are frequent and common. Mixed cropping is also practised with a view to achieve multiple requirements of food, feed and fibre. Mixed cropping is especially important for fodder crops where it can provide enormous quantities of feedstuff for supporting sustainable livestock production. In mixed cropping, the time of sowing of all the crops is almost the same. However, they may mature at a time (wheat + gram, wheat + barley or wheat + mustard) or they may mature at different times (redgram + jowar, sesame + greengram, wheat + *toria*, or bajra + groundnut).

Types of mixed cropping

Based on method of sowing and objective of cropping, mixed cropping can be classified into several groups.

Mixed cropping: The seeds of different crops are mixed together and then sown either in lines or broadcasted. The system is not scientific. Agricultural operations and harvesting of the crops become difficult.

Guard crops: The main crops is grown in the centre, surrounded by hardy or thorny crops such as safflower around pea or wheat, mesta around sugarcane, jowar around maize, etc. with a view to provide protection to the main crop.

Augmenting crops: When sub crops are sown to supplement the yield of the main crop, the sub crops are termed as augmenting crops. Mustard is sown on the borders of wheat fields to harvest mustard leaves for greens during the initial stages.

Multi-storeyed cropping or multi-tier cropping: Growing plants of different height in the same field at the same time is termed as multi-storeyed cropping. It is mostly practised in orchards and plantation crops for maximum use of solar energy even under high planting density. Examples of multi-tier cropping are *Eucalyptus* + papaya + berseem, sugarcane + potato + onion, sugarcane + *toria* + potato, and coconut + pineapple + turmeric/ginger.

Principles of mixed cropping

1. All crops in a mixed cropping system do not fail under adverse climate conditions.
2. Crops being attacked by similar insect pests and diseases should not be sown together.
3. The different types of crops are grown with an objective of fulfilling the daily need for cereals, pulses and oilseeds of a farm family.
4. The companion or subsidiary crops should not affect the growth of the main crop.
5. The subsidiary crops should mature earlier or later than of the main crop.
6. One of the crops should preferably be a legume.
7. The crops in a mixed cropping system should have different growth habits and nutrient requirements.
8. The crops should have different rooting depths and ramification.
9. The crops should not be very exacting in climatic requirements.

Advantages of mixed cropping

1. Mixed cropping ensures efficient utilisation of land, labour, equipment and other resources.
2. It utilises available space and nutrients to maximum extent possible.
3. It provides safe guard against hazards of diseases and pests.
4. It also reduces the effects of adverse weather conditions for the farmer as planting and harvest is at different times of the season.
5. Mixed cropping secures daily requirements like cereals, pulses, oilseeds, etc. to a farm family.
6. It improves yield. According to some scientific basis there is 10-25% increase in the yield in mixed cropping than monocropping.
7. It gives balanced cattle feed with both legume and grass fodders.
8. Costs of input decrease as compared to individual crop growing cost.

Limitations of mixed cropping

1. Some crops are very specific in the type of soil they need for maximum profits.
2. Crops for mixed cropping if not chosen properly have chances of competition between the crops for nutrients.
3. It also may reduce the fertility of the soil as more than one crop is grown at a time in the same piece of land.

Besides these, all the limitations for intercropping are also applicable for mixed cropping.

Table 3.1. Difference between intercropping and mixed cropping

Intercropping	Mixed cropping
1. The main object is to utilise the space left between two rows of main crop.	1. The main object is to get at least one crop under favourable conditions.
2. More emphasis is given to the main crop.	2. All crops are cared equally.
3. There is no competition between both crops.	3. There is competition between/among the crops grown.
4. Intercrops are of short duration and are harvested much earlier than main.	4. Usually the crops are almost of the same duration.
5. Sowing time may be same or different.	5. It is same for all crops.
6. Crops are sown in different rows.	6. Either sown in rows or mixed without considering the population of the crops.

3.2.2.3 Sequential cropping

Sequential cropping refers to growing crops in sequence within a crop year, one crop being sown after the harvest of the other. Here crop intensification is only in the time dimension without any spatial intensification like intercropping. One has to manage only one crop at a time in the same field. Intensification in sequential multiple cropping through introduction of non-conventional crops/short duration crop cultivars and intensive input management, is a common way of increasing land use efficiency especially in irrigated ecosystems (Gangwar et al., 2012). Irrigation facility, availability of short duration, photo-insensitive and thermo-insensitive high yielding varieties and other production inputs should be ensured for realising the maximum benefits from the sequence cropping.

Types of sequential cropping

Double cropping: Cultivation of two crops in succession on a piece of land in a year is called double cropping. Growing rice-wheat, rice-groundnut or red gram-maize in a year in sequence are the examples of double cropping.

Triple cropping: It refers to cultivation of three crops in succession on a piece of land in a year. Growing rice-maize-greengram, rice-potato-okra or red gram-maize-cowpea in a year in sequence are the examples of triple cropping.

Quadruple cropping: It refers to cultivation of four crops in succession on a piece of land in a year. Growing rice–fodder maize–greengram–greens, rice–potato–greens–okra or greengram–fodder maize–fodder cowpea–okra in a year in sequence are the examples of quadruple cropping.

Ratoon cropping or ratooning: It refers to raising a crop with regrowth coming out of roots or stalks after the harvest of the crop. Here, more than one harvest is done from one sowing/planting. Thus, ratooning consists of allowing stubbles of the original crop after harvesting and to raise another crop. In sugarcane 3-4 ratoon crops can be taken, but here a cropping year is not the time limit as it is a 10-18 months duration crop. Seasonal vegetables like okra, chilli, etc. can be taken as ratoon crops from the stubbles in the following season within a cropping year.

Gangwar and Singh (2011) have documented the efficient alternative cropping systems for various agroclimatic zones of the country (table 3.2).

Table 3.2. Identified high productive cropping systems (sequential cropping) for selected locations.

Locations	Prevailing cropping sequence		High productive alternative system	
	System	System rice equivalent yield (t/ha)	System	System rice equivalent yield (t/ha)
Jammu, J&K	Rice-wheat	11.3	Rice-marigold-french bean	30.1
			Rice-potato-onion	29.5
Ludhiana, Punjab	Rice-wheat	13.2	Maize-potato-onion	27.9
			Groundnut-potato-bajra (fodder)	23.3
Modipuram, U. P.	Rice-wheat	12.9	Maize-potato-sunflower`	24.2
			Rice-wheat-greengram	15.9
Sabour, Bihar	Rice-wheat	11.0	Rice-potato-onion	29.0
			Rice-wheat-maize	15.7
Bhubaneswar, Odhisha	Rice-rice	6.7	Rice-maize-cowpea	17.4
			Rice-maize-greengram	14.8
Coimbatore, Tamil Nadu	Cotton-sorghum -finger millet	4.1	Beet root-greengram-maize+cowpea	7.1
			Chillies+onion-Sunhemp -okra+coriander	6.6
Thanjavur, Tamil Nadu	Rice-rice-sesame	13.7	Rice-rice-brinjal	18.3
			Direct seeded rice-rice-maize + blackgram	17.4
S.K. Nagar, Gujarat	Groundnut-wheat-fallow	4.1	Groundnut-wheat-sesame	7.0
			Groundnut-onion-greengram	5.0
Bengaluru, Karnataka	Hybrid cotton -sunflower	7.0	Maize-groundnut	12.2
			Maize-sunhemp` -sunflower	11.3
Hyderabad, Telangana	Rice-rice	7.9	Maize-onion	12.3
			Maize-tomato	12.1
Mean		9.2		16.9

Source: Gangwar and Singh, 2011

Advantages of sequential cropping

1. It increases the overall productivity of the system.
2. It also influences the soil conditions favourably.
3. It ensures efficient utilisation of available resources like nutrients, water and solar energy.
4. Positive residual influence of previous crop with respect to plant nutrients on succeeding crops is realised.

Disadvantages of sequential cropping

1. Pests and diseases from a crop may be carried over to the succeeding crop.
2. Weed population and species may sometimes be shifted from one crop to second crop.
3. The growth of second crop may be adversely affected by allelopathic effect of first crop. Sunflower has allelopathic effect on succeeding crops like mustard and soybean by inhibiting germination of their seeds.

Table 3.3. Difference between intercropping and sequential cropping

Intercropping	Sequential cropping
1. Growing of two or more crops simultaneously on the same piece of land in a cropping season.	2. Growing of two or more crops in sequence on the same piece of land in a farming year.
2. Intercropping is intensive cropping mainly in space dimension.	2. Sequential cropping is intensive cropping mainly in time dimension.

3.2.2.4 Crop rotation

Crop rotation refers to growing of different crops alternatively on the same piece of land in a definite sequence or process of growing different crops in succession on a piece of land in a specific period of time with an objective to get maximum profit from least investment without impairing the soil fertility. This may also be defined as the repetitive cultivation of an ordered succession of crops (or crops and fallow) on the same land and one cycle may take one or more years to complete.

Principles of crop rotation

1. Crops best suited to the climatic conditions and soil type should be selected.
2. The crops with taproots should be followed by those with fibrous root system. This helps in proper and uniform use of nutrients from the soil and crops do not compete with each other for uptake of nutrients.

3. The leguminous crops should be grown after non-leguminous crops. Legumes fix atmospheric nitrogen in the soil and add more organic matter to the soil in the form of huge quantity of root and foliage.
4. More exhaustive crops should be followed by less exhaustive crops. Exhaustive crops like sugarcane, maize, etc. should be followed by low inputs demanding crops like oilseeds and pulses.
5. The crop of the same family should not be grown in succession because they act like alternate hosts for pests and diseases.
6. The crop rotation should provide maximum employment to the farm family, farm machineries and equipments available.
7. Selection of the crops should be based on local demand, storage, transport and marketing facility.
8. Grower's preference and his socioeconomic conditions and resource mobilising power should also be considered.
9. The selection of the crops should be based on farmer's knowledge and skill.
10. The latest production technology of the crops should readily be available.
11. On slopes which are prone to soil erosion, alternately erosion-promoting crops like millets or other row crops and erosion-resisting crops like legumes should be adopted.
12. Under dryland farming or partially irrigated areas, the crops like pulses, sunflower which can tolerate drought spells to some extent should be selected. Similarly in low lying and flood prone areas the crops like rice and jute which can tolerate water stagnation should be selected.

Benefits of crop rotation

1. Soil fertility is restored by fixing atmospheric nitrogen by integrating legumes in a rotation.
2. Proper selection of crops in a rotation encourages soil microbial activity and improves physico-chemical properties of the soil.
3. The succeeding crop utilises the residual nutrients.
4. The soil is also protected from erosion, salinity and acidity.
5. It controls pests and diseases if dissimilar crops are taken in succession.
6. It also restricts the weeds growth.
7. Farmers get better price for their produce because of its higher demand in the locality or in the market.
8. The family labour, power, equipment and machineries are well employed.
9. Proper utilisation of all the resources and inputs could be made.

Limitations of crop rotation

1. Specialisation in many crops is not possible always.
2. Requirement of equipments and machineries varies from crop to crop.
3. Allelopathic effect of preceding crop affects the succeeding crop.
4. Similar types of crops serve as alternate hosts for pests and diseases.

3.3 PLANT INTERACTIONS IN MULTIPLE CROPPING

In intensive cropping, when crops are grown in association (intercropping or sequential cropping) interaction between different component crop species occurs, which is essentially a response of one species to the environment as modified by presence of another species. This phenomenon is commonly referred as interference or interaction.

3.3.1 Nature of Interactions

The nature of interactions between two components can be described on the basis of observable net effect of one component on another in a system. The basic principle deciding the nature of interactions depends upon the ability of different components to capture and use the most limiting essential growth resources effectively. The capture of the limiting resource (e.g. light, water or nutrients) depends on the number, surface area, distribution and effectiveness of the individual elements within the canopy or root system of the species or mixture of species involved.

3.3.1.1 Complementary interaction

In a system if the component crops help each other, by creating favourable conditions for their growth in such a way that the system provides a greater yield than the yield of their corresponding sole crops, then the interaction between the components is said to be complementary in nature. The complementary interaction is of two types, viz. spatial and temporal. When complementarity is achieved simply by varying the proportion of the desired species, it is known as spatial complementary interaction. In this, the geometry of the system is altered by changing the spacing and density of the components. Therefore, under complementary association the proportion of each species can be modified to suit one's requirements without any sacrifice. When two species complement each other, the capture of the major limiting growth resources by the mixture always exceeds that by the corresponding sole stands. The two components also complement each other through more effective use of growth resources by utilising them at specific times due to difference in growth pattern and resource requirement. There are substantial opportunities for temporal complementarities if species make their major demands on available resources at different times, thereby reducing the possibility of competition.

2.3.1.2 Supplementary interaction

If the two components interact in such a way that the yield of one component exceeds the yield corresponding to its sole crop without affecting the yield of other component, the interaction is known to be supplementary in nature. Therefore, there will be an additional yield from one of the crops; thus the overall yield of the system will be more than that of sole crops of both the components. In this interaction one component is maximising its resource utilisation without limiting it for the other component.

2.3.1.3 Competitive interaction

In this system the components interact in such a way that the increase in the yield of one component leads to decrease in the yield of other component due to competitive interaction. One species may have greater ability to use the limiting factor and will gain at the expense of the other. The intensity of competition is the greatest when requirements are similar, and the growth and development proceed synchronously for the component crops. Thus, there is extensive overlap between species in their resource requirements, resulting in a severe competition. To reduce the competition in a system, the components should be selected in such a manner that they can utilise the finite resources (water, nutrients, solar energy, etc.) at different times and show different growth and development phases from each other.

3.3.2 Factors Affecting Interactions

The interactive relations among the component crops of any system are affected by many factors such as choice of species, population of species, site factors, management practices, etc. Many of these factors can be manipulated for better production from the system.

Species of crops: The growth pattern of crop plants varies from species to species and thus, affects the interactions with the associated components. Some crops may perform better in association with a particular crop species, whereas the yield of other crops may reduce with the same crop species.

Plant population: Yield per plant linearly correlated with the available space. In close planting, leaf surface per plant and unit area is reduced. Closely planted crops have a uniform root spread. Widely spaced crops have a circular root distribution which is beneficial to crops. High plant density leads to seedling mortality. Build-up and spread of pest and diseases are more in high plant density without proper management.

Site factors: Variations in climatic (rainfall, temperature, humidity, etc.), edaphic (physical, biological and chemical properties) and physiographic features of an area affect the plant growth which leads to variation in the interactive relationship of component species in system.

Management practices: The interactive relationship between or among the different crop species also varies with management practices adopted for the system. Through various management practices the interactions among the species in a system can be manipulated.

3.3.3 Interactions in Intercropping System

In an intercropping system, always there is an interaction between different component crops. This interaction may be positive (beneficial) or negative (harmful). These positive or negative interactions can be direct or indirect or both. Both positive and negative interactions may further be divided into above or below ground interaction.

The main effects of positive interactions in intercropping systems are increased productivity, improved soil fertility, nutrient cycling, soil conservation, water conservation, weed control and microclimate improvements. Similarly, the effects of negative interactions include light competition, nutrient competition, water competition, pests and diseases and allelopathy. However, these interaction processes are interdependent and the manifestation of their effects is influenced by the environmental conditions.

3.3.3.1 Positive interactions

Increased productivity: The biomass productivity of an intercropping system is generally greater than that of a monocropping system. The basis for the potentially higher productivity is due to the capture of more growth resources (e.g. light, nutrient and water) or due to improved soil fertility through biological nitrogen fixation by the legumes. The presence of two or more species of different growth habits leads to a net effect of complex interactions resulting into microclimate amelioration, thus affecting transpiration rate, photosynthesis and energy balance of the associated crops. All these factors may translate into increased productivity from the system.

Improved soil fertility: The potential for micro-site enrichment by some crop species is an extremely important aspect of intercropping when a legume is integrated. In alley cropping use of fast growing, nitrogen-fixing trees (e.g. *Leucaena leucocephala* and *Gliricidia sepium*) can substantially increase the soil fertility. A major feature of this system is to produce a large quantity of biomass which enriches the soil. Presence of rhizosphere microflora and mycorhiza associated with one of the crops may lead to mobilisation and availability of nutrients which may benefit the associated crops also. Legumes in an intercropping system supply part of N fixed by symbiosis to non-legumes like in case of maize+cowpea and rice+groundnut.

Nutrient cycling: Soil nutrients are considered among the least resilient components of sustainability. A fundamental principle of sustainability therefore is to return to the soil the nutrients removed from it through harvests, runoff, erosion, leaching, etc. In intercropping systems, these nutrient loss pathways are minimised. The legume components provide additional nutrient inputs through biological nitrogen fixation. If component crops are of different root depths, they can capture nutrients from different layers of soil. The nutrients captured from deeper layers by the deep-rooted crops are transferred to the soil through litter (leaves, crop residues, root debris, etc.) decomposition.

Soil conservation: Intercropping systems usually have more plant population and cover the soil surface for greater period in a year as compared to monocropping. This is highly effective in controlling soil erosion. The thick vegetation provides a semipermeable barrier to surface movement of water and reduces the impact of rain drops on the soil and minimises splash.

Water conservation: Thick vegetation increases the infiltration of rain water while simultaneously reduces evaporation from the soil.

Weed control: Another potentially positive interaction in intercropping systems is related to weeds. Higher plant population causes shade on the soil surface and shade suppresses light-demanding weeds. Weed biomass yield is positively correlated with available solar energy.

Annidation: Complementary interactions which occur both in space and time are referred to annidation. Certain crops require less light intensity and high relative humidity. Such an altered microclimate is provided when such crops are grown in between tall growing components in an intercropping system. This is annidation in space. Growing turmeric, ginger, pineapple or black pepper in coconut plantation uses this principle. The component crops utilising resources from different soil layers in redgram and rice or groundnut intercropping system is an example of annidation in space. When two crops of widely varying duration are planted, their peak demands for light and nutrients are likely to occur at different periods, thus reducing competition. This is annidation in time. When early maturing crop is harvested it becomes favourable for late maturing crop in intercropping systems like sorghum+redgram, groundnut+redgram, maize+greengram, etc.

3.3.3.2 Negative interactions

As all the crops in an intercropping system utilise the same reserves of growth resources, negative interactions are likely to occur in every crop plant association. The major yield decreasing effects arise from competition for light, water, and nutrients, as well as from interactions through allelopathy.

Competition for light: In intercropping, shading by the taller crop plants reduces light intensity at the short height crop and this is the most limiting factor in many situations, particularly those with relatively fertile soils and adequate water availability. The relative importance of light will decrease in semiarid conditions as well as on sites with low fertility soils. Since crops differ in their responses to poor nutrition, competition for light may either be reduced or amplified by a shortage of nutrients. The growth duration of the component crops play a major role to minimise competition for light. The difference in maturity of component crops should be at least 30 days to get the maximum benefit from an intercropping system. Proper choice of crops and genotypes, adjustment of population density, judicious proportion of each component in the mixture and correct planting pattern are the ways of increase the light use efficiency. The best way to orient crop rows in the northern hemisphere is north to south. This gives the most sun exposure and allows for ample air circulation. When crops are planted east to west, the rows tend to shade each other.

Competition for nutrients: If the peak demand period for the nutrients coincides for all the component crops in an intercropping system there is a competition for nutrients. The effect of nutrient competition is more severe for the crops when the root systems of different crops confined to the same soil horizon and availability of plant nutrients in the rhizosphere is limited. Competition for nutrients also depends upon the root spacing of dominated and aggressive crops.

Competition for water: Competition for water is likely to occur in intercropping systems at some period of time except in areas with well distributed rainfall or continuous supply of irrigation water. The competition depends on the severity of the drought and the drought tolerance of the crop plants. It plays a major role in the productivity of the systems despite the use of drought-tolerant and drought-adapted plants, especially in dry areas. Competition for water also depends on the method of irrigation, earliness of water demand, root extension (lateral and vertical growth) of the component crops.

Microclimatic modification for pests/diseases: Bacterial and fungal diseases may increase in shaded, more humid environments in an intercropping system. Reduced temperature and humidity fluctuations under shade can also have a favourable effect on spread of pests and diseases. One crop can be a host of insect pests and diseases for another crop of the system.

Allelopathic interaction: Allelopathy refers to the inhibition of growth of one plant by chemical compounds (allelochemicals) that are released into the soil from neighbouring plants. Some crops may be unsuitable as intercrops because they secrete toxins into soil which will adversely affect the associated crops. Allele chemicals are reported to be present in practically all plant tissues, including leaves, flowers, fruits, stems, roots, rhizomes and seeds. The effects of these chemicals are dependent mainly upon the concentration as well as the combination in which one or more of these substances is released into the environment. The organic compounds released this way are often phytotoxins. Such toxic chemicals may be injurious to microbes and even to the seedlings of those plants releasing them. The effects of the chemicals may result in complete inhibition of growth or retarded growth. Allelopathic compounds may be released into the environment by volatilisation, leaching from living or dead tissues, exudation from roots and decay of plant tissues. The mechanism of action of allelochemicals is diverse and includes.

1. Inhibition of cell division and elongation
2. Inhibition of action of gibberellins or indole acetic acid that induces growth
3. Reduction of mineral uptake
4. Retardation of photosynthesis
5. Inhibition or stimulation of respiration
6. Inhibition or stimulation of stomatal opening
7. Inhibition of protein synthesis and changes in lipid and organic acid metabolism

3.3.4 Interactions in Sequential Cropping System

In sequential cropping, the preceeding crop has considerable influence on the succeeding crop mainly by changes in soil conditions, presence of allelopathic chemicals, shift in weed flora and carryover effects of fertilisers, pests and diseases.

Soil conditions: A few problems are encountered for land preparation in sequential cropping system. The time available for seedbed preparation is less in high intensity cropping system. Due to the effect of the previous crop, the field may not be in proper condition to carry out field operations. Field preparation becomes difficult after transplanted rice because the soil structure is destroyed due to puddling. The establishment of a pulse crop after rice is also difficult. Delay in sowing or planting is the most common problem in sequential cropping systems because the field preparation and planting of the succeeding crop is delayed. Short duration varieties of crops are selected to fit well in sequential cropping systems. Proto- and thermo-insensitive varieties are essential for successful sequence cropping system.

Allelopathic interaction: Crops like sorghum and sunflower leave toxic chemicals in the soil which do not allow germination of subsequent crops. Roots of cucumber, leaves of *Eucalyptus*, decomposing residues of sunflower are known to produce allele chemicals affecting the growth of succeeding crops.

Weed flora: Weed number and species differ in the succeeding crop due to the effect of the previous crop. It is reported that wheat crop that follows rice suffers from high density of *Phalaris minor*. Weed management practices are followed for the entire system as a unit instead of considering weed problem of individual crops. Herbicides applied to the previous crop may be toxic to the succeeding crop. Higher dose of atrazine applied to sorghum crop affects germination of succeeding pulse crop. Herbicide and its dose for the preceding crop should, therefore, be recommended with the consideration of succeeding crop.

Soil fertility: The previous leguminous crop leaves considerable amount of nitrogen in the soil for the succeeding crop. The legumes also add a huge quantity of organic matter rich in nitrogen in the soil in the form of root residues and leaf litter. P applied to the previous crop is also available for the succeeding crop.

Pests and diseases: The infestation of pests and diseases are more in sequence cropping system due to continuous cropping. The pests and diseases in crop stubbles and other residues of the previous crop may infect the subsequent crop. However, carryover effects of insecticides are not observed.

3.4 CROP DIVERSIFICATION

Crop diversification can be considered as an attempt to increase the diversity of crops through, e.g. crop rotation, multiple cropping or intercropping, compared to specialised farming with the aim to improve the productivity, stability and delivery of ecosystem services. It can be one measure to develop more sustainable production systems, develop value-chains for minor crops and contribute to socioeconomic benefits. Crop diversification practices can include higher crop diversity, more diverse crop rotations, mixed cropping, cultivation of grain legumes in otherwise cereal dominated systems and regionally adapted varieties or variety mixtures. Crop diversification and/or additional diversification measures like variation of seeding time or changing cropping patterns have the potential to lead to higher and more stable yields, increase profitability and lead to greater resilience of agroecosystems in the long term. These practices have the potential to make cropping systems more diverse in space, time and genetics. Consequences of diversification are temporal shifts and ranges of phenological stages (relevant for biodiversity and adaptation to climate change), more frequent or continuous soil cover and more diverse management strategies, i.e. 'tillage', 'sowing dates', 'fertilisation', 'irrigation', 'harvesting' and also reducing labour peaks and economic risk (Hufnagel et al., 2020).

In Indian context, crop diversification is a demand driven, need based situation specific and national goal seeking continuous and dynamic concept and involves spatial, temporal, value addition and resource complementary approaches. Diversified farms are usually more economically and ecologically resilient. While monoculture farming has advantages in terms of efficiency and ease of management, the loss of the crop in any one year could put a farm out of business and/or seriously disrupt the stability of a community dependent on that crop. By growing a variety of crops, farmers spread economic risk and are less susceptible to the radical price fluctuations associated with changes in supply and demand.

With the advent of modern agricultural technology, especially during the period of the 'green revolution' in the late sixties and early seventies of the last century, there is a continuous surge for diversified agriculture in terms of crops, primarily on economic considerations. The crop pattern changes, however, are the outcome of the interactive effect of many factors which can be broadly categorised into the following five groups (Hazra, 2001).

1. Resource related factors covering irrigation, rainfall and soil fertility.
2. Technology related factors covering not only seed, fertiliser, and water technologies but also those related to marketing, storage and processing.
3. Household related factors covering food and fodder self-sufficiency requirement as well as investment capacity.
4. Price related factors covering output and input prices as well as trade policies and other economic policies that affect these prices either directly or indirectly.
5. Institutional and infrastructure related factors covering farm size and tenancy arrangements, research, extension and marketing systems and government regulatory policies.

All these five factors are interrelated. The adoption of crop technologies is influenced not only by resource related factors but also by institutional and infrastructure factors. Similarly, government policies, both supportive and regulatory in nature, affect both the input and output prices. Likewise, special government programmes also affect area allocation and crop composition. Although the factors that influence the area allocation decision of farmers are all important, they obviously differ in terms of the relative importance both across farm groups and resource regions. While factors such as food and fodder self-sufficiency, farm size, and investment constraints are important in influencing the area allocation pattern among smaller farms, larger farmers with an ability to circumvent resources constraints usually go more by economic considerations based on relative crop prices than by other non-economic considerations. Economic factors play a relatively stronger role in influencing the crop pattern in areas with a better irrigation and infrastructure potential. However, the relative importance of these factors change over time. From a much generalised perspective, Indian agriculture is increasingly getting influenced more and more by economic factors. This need not be surprising because irrigation expansion, infrastructure development, penetration of rural markets, development and spread of short duration crop varieties and drought resistant crop technologies have all contributed to minimising the role of non-economic factors in crop choice of even small farmers (Hazra, 2001).

3.4.1 Approaches to Crop Diversification

There are several approaches to crop diversification for achieving the sustainability in agriculture.

1. The overall productivity of a farm may be increased by crop intensification with addition of more number of crops (intercropping, mixed cropping, sequential cropping, multitier cropping, etc.) to the existing crops or cropping systems.
2. The crops which are less suitable or productive under an agroecological region may be substituted with more suitable alternate crops. More remunerative crops but with high risk to a drought prone area should be substituted with low risk crops like short duration pulses and drought resistant oilseed crops. Enough elasticity should be kept in rotation so that if pest or diseases destroys a crop, another crop can be substituted.
3. Selection of crops should be based on land type and situation. Erosion resisting crops like legumes and fodder grasses that cover the land fast should be taken on sloppy lands which are prone to soil erosion.
4. Fertile and well-drained land should be utilised for important food-crop rotation, and less fertile land may be diverted for restorative crops or soil improving crops like legumes and crops with high foliage. Similarly, tolerant crops should be selected for problem soils like acidic, saline or alkali soils.
5. Conventional crops sometimes fail to provide stability of production over a period of time due to abiotic stresses. Tree-based cropping systems like alley cropping, agrisilvicultural or silvipastoral or silvihortipatoral systems are more sustainable in dry areas.

6. Marginal lands are permanently disadvantaged in terms of steep slopes, poor drainage, poor soil depth and harsh environments. Such lands can be developed into pastures or can put to use for tree farming. Ley farming, i.e. growing of grass or legumes in rotation with grain crops may also practised in these lands. Grasses improve soil structure while the legumes enrich soil nitrogen status. Grasses such as *Cenchrus* sp., *Chloris* sp., *Panicum* sp. can also be grown in mixture with perennial pasture legumes.
7. Degraded lands may be used with suitable tree plantation. Legume fodder trees should be preferred.
8. Low water requiring crops should replace the high water-demanding crops in dryland areas and water scarce regions.
9. Two or three crops in sequence are usually taken in a year under irrigated conditions. However, a dry crop should be included in the rotation to avoid damage to the soil due to continuous irrigation. Rice–rice system should be discouraged.
10. The cropping intensity of a region varies with the amount of annual rainfall received. Intercropping or mixed cropping with the companion crops having varying peak moisture-demanding period may be encouraged in the areas where moisture may be enough to produce one crop but not enough to produce two in a sequence.
11. Root crops are well adapted to the humid lowland than non-rice cereals and legumes since high rainfall and humidity often adversely affect reproduction, ripening, drying and storage and increase pest and disease problems of cereals and grain legumes. In areas where drainage is not a problem, vegetables may be sustainable.
12. Low yielding varieties can be substituted with high yielding varieties to sustain the production. Always new high yielding varieties or hybrids which are resistant or tolerant to biotic and abiotic stresses should be given priority.
13. All the principles of crop rotation should be strictly followed to have sustainable production over the years.
14. Selection of the crops should always be based on soil, climate and market demand.

3.4.2 Advantages of an Ideal Crop Diversification

An ideal crop diversification brings sustainability in agriculture with the following major advantages.

1. Inclusion of crops of different feeding zone and nutrient requirement could maintain the better balance of nutrient in soil. Growing crops of different root depths avoids continuous depletion of nutrients form same depth.
2. Crops requiring high irrigation when followed by crops requiring light irrigation, the overall water use efficiency and water productivity of the system is increased.
3. It improves soil structure, percolation and reduces chances of creation of hard-pan in subsoil and also reduces soil erosion.
4. There is an overall increase in yield of crops mainly due to maintenance of physical and chemical properties of soil. Soil fertility is restored by maintaining more organic matter, encouraging microbial activity and biological cycles and protecting the soil from erosion, salinity and acidity.
5. It helps in controlling insects, pests and soil borne diseases. It also controls weeds.
6. Diversification of crops reduces risk of financial loss due to unfavourable conditions.
7. The family needs of feed, food, fuel, fibre, etc. are fulfilled.
8. It facilitates even distribution of labour.
9. Farmers get a better price for their produce due to higher demand in local market. So there is regular flow of income over years.

3.4.3 Constraints in Crop Diversification

The major problems and constraints in crop diversification in India are primarily due to the following reasons with varied degrees of influence (Hazra, 2001).

1. Over 117 m ha (63%) of the cropped area in the country is completely dependent on rainfall.
2. Suboptimal and overuse of resources like land and water resources, causing a negative impact on the environment and sustainability of agriculture.
3. Inadequate supply of seeds and plants of improved cultivars.
4. Fragmentation of land holding restricts modernisation and mechanisation of agriculture.
5. Poor basic infrastructure like rural roads, power, transport, communications, etc.

6. Inadequate post-harvest technologies and inadequate infrastructure for post-harvest handling of perishable horticultural produce.
7. Very weak agro-based industry.
8. Weak research - extension - farmer linkages.
9. Inadequately trained human resources together with persistent and large-scale illiteracy amongst farmers.
10. Host of diseases and pests affecting most crop plants.
11. Poor database for horticultural crops.
12. Decreased investments in the agricultural sector over the years.

3.4.4 Alternate Cropping Systems

Several alternate cropping systems have been identified and recommended by All India Coordinated Research Project on Cropping Systems for various agroclimatic regions of India. The yield increase over existing systems with alternate efficient cropping systems was found to be 40 to more than 300% in various agroclimatic zones.

Table 3.4. Alternate cropping systems under various agroclimatic zones

Agroclimatic zones	Location	Existing system	Alternative systems	% increase over existing system
Western Himalayas	Dhansaur	Rice–wheat	Rice–potato–onion	230.7
		Maize–wheat	Maize–potato–onion	338.5
	Pantnagar	Rice–wheat	Rice–potato–vegetable cowpea	123.3
Eastern Himalayas	Karimganj	Rice–potato	Rice–rajmah	42.5
Lower Gangetic Plains	Kakdwip	Rice–greengram	Rice–okra	106.9
Middle Gangetic Plains	Patna	Rice–wheat	Rice–wheat–greengram	59.8
Upper Gangetic Plains	Saini	Rice–wheat	Sesame–pea	97.8
Eastern Plateau and Hills	Kawardha	Soybean–gram	Soybean–tomato–cowpea	244
	Dhenkanal	Rice–greengram	Rice–tomato	116.9
	Gondia	Rice–wheat	Rice–wheat–cowpea	76.3
Central Plateau and Hills	Udaipur	Maize–wheat	Maize–wheat–okra	359.6
Western Plateau and Hills	Aurangabad	Soybean + pigeonpea	Soybean–gram–fodder maize	208.4
	Ahmednagar	Soybean–wheat	Soybean–onion	238.9
Southern Plateau and Hills	Warrangal	Rice–rice	Rice–maize	39.9
	Bangalore	Rice–finger miller	Rice–brinjal	250.7
	Paiyur	Rice–rice	Rice–tomato	42.2
East Coast Plains and Hills	Kendrapara	Rice–greengram	Rice–bittergourd	88.9
West Coast Plains and Ghats	Thiruvalla	Rice–rice	Rice–rice–okra	243.5
	Roha	Rice–cowpea	Rice-maize	105.7
Gujarat Plains and Hills	Thasara	Tobacco	Tobacco–fodder bajra	169.6
		Bajra–wheat	Bajra–lucerne (seed)	129.5
	Deesa	Pearlmillet-mustard	Greengram–fennel	55.5
		Castor	Fennel–fodder pearlmillet	41.1

Source: Gangwar et al., 2012

4

Efficiency Evaluation of Cropping Systems

The objective of any cropping system is efficient utilisation of all resources such as land, water and solar radiation, maintaining stability in production and obtaining higher net returns. The efficiency is measured by the quantity of produce obtained per unit resource used in a given time. However, widespread occurrence of second-generation problems, such as over-mining of soil nutrients, decline in productivity, reduction in profitability, lowering of groundwater tables and build-up of pests including weeds, diseases and insects has been reported during post-green revolution era in most of the intensively cultivated, high-productivity, cereal based production systems, which are threatening their sustainability. Studies carried out under All India Coordinated Research Project on Cropping Systems have resulted in identification of appropriate duration of varieties for some popular crop sequences for different regions of the country. Several alternative cropping systems with high land use efficiency and monetary returns have also been identified for different agroecological and farming situations of the country. These multiple cropping system included sequential as well as intercropping systems, including several variants such as mixed cropping, relay cropping, alley cropping, parallel multiple cropping, multistoried cropping, etc.

4.1 EFFICIENT CROPPING SYSTEMS

Efficient cropping systems depend on available farm resources and production technologies. The farm resources include land, labour, water, capital and infrastructure. When land is limited intensive cropping is practised to fully utilise the other available resources like water and labour. When sufficient and cheap labour is available, vegetable crops may be included in the cropping systems as these are labour-intensive. Capital-intensive crops like sugarcane, turmeric and short duration fruit crops are included in the cropping systems when capital is not a constraint. Double, triple and quadruple cropping systems are adopted with availability of adequate irrigation facility. When the farm enterprises include dairy, cropping system should have fodder crops.

4.1.1 Cropping Systems for Higher Nutrient Use Efficiency

Inclusion of legumes in the cropping system has been known since times immemorial. Legumes play a vital role in increasing indigenous nitrogen production in the field. These help in solubilising insoluble P in soil, improving the soil physical environment, increasing soil microbial activity and restoring organic matter. The carryover of N derived from legume grown, either in crop sequence or in intercropping system for succeeding crops, is also important. Thus, the scope for exploiting direct and residual fertility due to legumes has a great potential.

Sorghum, pearl millet, maize and castor are usually grown in dry lands and marginal and sub-marginal lands. Sorghum yield increased when sown after cowpea, green gram, and groundnut. Legumes like groundnut or cowpea provide an equivalent to 60 kg N/ha on the subsequent crop of pearl millet. Incorporation of whole plant of summer greengram or blackgram into soil (after picking pods) before transplanting rice is very much beneficial to rice-wheat system and up to 25% N can be saved in rice. Legumes with indeterminate growth are more efficient in nitrogen-fixation than determinate types. Fodder legumes in general are more potent in increasing the productivity of succeeding cereals. The carryover of N for succeeding crops may be 60-120 kg in berseem, 75 kg in cluster bean, 35-60 kg in fodder cowpea, 68 kg in chickpea, 55 kg in blackgram, 54-58 kg in groundnut, 50-51 kg in soybean, 50 kg in *Lathyrus*, and 36-42 kg per ha in redgram (Ghosh et al., 2007).

The major consideration in N management in intercropping systems is to quantify the 'direct transfer' of N from legume component to the non-legume component. However, their crop components have different requirement for nutrients. Cereals have less P but high N requirement, while legumes possess effective mechanism for symbiotic N fixation but have a high P requirement. Because of this, fertiliser use is considered complicated. N economy through intercropped legumes is to be correctly assessed in different intercropping systems. In legume and cereal intercropping system, the maize-pigeon pea is highly suitable with a minimum competition for nutrients, while in legume and legume intercropping system, pigeon pea-groundnut system is the most efficient one in terms of resource use-efficiency. Pigeon pea starts flowering after the maize has been harvested and its period of greatest nutrient demand occurs when the maize has already completed its growth cycle. However, in sorghum-pigeon pea system, grand growth period of both the crops falls exactly at the same time as a result of which tall cereal adversely affects the growth of the associated pigeon pea. Further, higher amounts of N applied to sorghum-pigeon pea system would result in better growth and yield of sorghum and consequently the growth and yield of pigeon pea will be adversely affected. Therefore, moderate quantity of N (25-

50 kg N/ha) should be applied to sorghum-pigeon pea system under dry land condition to achieve higher yield advantage (Stern, 1993).

4.1.2 Cropping Systems for Weed Management

Intermittent crop substitution/diversification in sequential cropping systems has been found to reduce some obnoxious weeds to a considerable extent, thereby reducing herbicide need to a great extent in areas where such weeds have assumed alarming proportions due to continuous adoption of a certain cropping system. Johnson grass (*Sorghum halepense*) becomes predominant weed in continuous maize cultivation but can be controlled by rotating with cotton. Similarly, a change from rice-wheat and rice-potato system to any other system, not involving rice in rainy reason, tends to reduce population of *Phalaris minor* in wheat considerably.

The crops that quickly form a shade canopy and are allelopathic in nature have an adverse impact on weeds sensitive to shade and can be adopted as weed control measures. Berseem (*Trifolium alexandrinum*) may be taken as a break crop successfully for reducing weed problem in continuous rice-wheat system without any monetary loss. Weed management through cover crops, intercrops, relay cropping are best strategies for cultural control of weeds.

4.1.3 Cropping Systems for Disease and Pest Management

Besides minimising herbicide use for weed control, insecticide/fungicide use can also be minimised to a considerable extent through cropping system approach. Sorghum earhead fly damage is extremely rare where pigeon pea is planted in alternate rows. Incidence of root rot of cotton caused by *Rhizoctonia solani* fungus is appreciably reduced by intercropping of moth bean (*Phaseolus aconitifolius*). The moth bean intercropping causes moderating effect on soil temperature increase which is unfavourable for the parasitic activity of fungus. It is also reported that coriander, garlic or fennel (*Foeniculum vulgare*) intercropped with autumn planted sugarcane prevents top borer attack in sugarcane and there is no need to apply insecticides for the control of top borer in sugarcane intercropping with these crops.

4.1.4 Cropping Systems for Higher Water Use Efficiency

Irrigation is a commonly used platform for crop intensification because it offers a point where the potentiality of other inputs is fully exploited. Making this sustainable intensification, more crop and profit per drop of water approach should be followed which is possible only through carefully designed cropping systems (Gangwar et al., 2012). Irrigation water is a costly and scarce resource and its availability for agriculture is expected to further go down due to increased demand for domestic and industrial uses. Though water use efficiency can be increased by genetic and environmental manipulation of the crops, it can also be increased by decreasing the evapotranspiration and other losses of water, such as conveyance, application, distribution and deep percolation. In cropping systems perspective, water use efficiency can be increased by identification of appropriate crop combinations in various systems. More remunerative and less water consuming crop rotations have been standardised at different locations in the country. Rice-mustard-sesame, rice-mustard-greengram and rice-potato-greengram rotations have been identified as more water efficient systems at Memari in West Bengal. At Kharagpur, rice-wheat, rice-mustard and rice-potato are more viable sequences under lesser water input. At Chiplima (Odisha) rice-mustard, rice-wheat-greengram and rice-potato-sesame rotations were identified as most water efficient systems. In Tarai conditions of Uttar Pradesh, under higher levels of irrigations rice-lentil and rice-wheat cropping systems were better while under lowest levels of *kharif* and *rabi* irrigations soybean-wheat proved the best sequence. Water productivity can be considerably increased through relay intercropping systems.

4.1.5 Cropping Systems in Context of Climate Change

If a cropping system becomes unsustainable due to any reason, it should be replaced by a new emerging system. A farmer is always interested in cropping systems that have a balance mix of higher biological productivity and economic returns with lesser risk and which possesses more stability over a period of time under fluctuating environmental situations. Indian agriculture is highly prone to the risks due to climate change, especially to drought, because two third of the agricultural land in India is rainfed, and even the irrigated system is largely dependent on monsoon. Flood is also a major problem in many parts of the country. Increase in temperature can reduce crop duration, increase crop respiration rates, alter photosynthesis process, affect the survival and distributions of pest populations and thus developing new equilibrium between crops and pests, hasten nutrient mineralisation in soils, decrease fertiliser use efficiencies, and increase evapotranspiration.

Development of new crop varieties with higher yield potential and resistant to multiple stresses (drought, flood, salinity) will be the key to maintain yield stability. On-farm water conservation techniques, micro-irrigation systems for better water use efficiency and selection of appropriate crop has to be promoted. Adjustment of planting dates to minimise the effect of high temperature induced spikelet sterility can be used to reduce yield instability so that the flowering period does not coincide with the hottest period. The conventional puddled transplanted rice and intensively tilled wheat cropping system should be changed to other cropping systems such as maize-wheat, pulse-wheat, maize-pulse, oilseed-wheat and direct-seeded rice-wheat. These systems have less demand for water and nutrient and use resources more efficiently.

4.2 EFFICIENCY ASSESSMENT OF CROPPING SYSTEMS

Several indices have been developed and are being used for evaluating system productivity, input use efficiency and economic efficiency of multiple cropping systems.

4.2.1 Assessment of Land Use

One of the major objectives of cropping systems is to use available resources efficiently. There are several indices to compare the efficiency of different cropping systems in terms of land use.

Multiple cropping index (MCI)

This measures the sum of area planted to different crops and harvested in single year divided by the total cultivated area times 100.

$$MCI = \frac{\sum_{i=1}^{n} ai}{A} \times 100$$

where, n = Total number of crops

ai = Area occupied by i^{th} crop planted and harvested within one year

A = Total cultivated land area available

This index gives the percent utilisation of the land area and is variant of the cropping intensity which does not account for the cultivars grown in mixed or intercropping systems.

Land utilisation index (LUI)

Land utilisation index has been defined as the number of days during which the crops occupied the land during a year divided by 365. This index can be expressed as a fraction or as a percentage. It gives only an idea how the land has been put to use. If the index is 1, the land has not been left fallow and if more than 1, it tells the specifications of intercropping and relay cropping, etc.

Cultivated land utilisation index (CLUI)

This index is calculated by summing the products of land area planted to each crop, multiplied by the actual duration of that crop and divided by the total cultivated land area available during 365 days.

$$CLUI = \frac{\sum_{i=1}^{n} aidi}{A \times 365}$$

where, n = Total number of crops

ai = Area occupied by i^{th} crop

di = Days that i^{th} crop occupied ai

A = Total cultivated land area available during the 365 days period

Cropping intensity (CI)

Cropping intensity refers to raising of a number of crops from the same field during one agricultural year. It is the ratio of effective crop area harvested to the physical area.

$$\text{Cropping intensity} = \frac{\text{Gross cropped area}}{\text{Net sown area}} \times 100$$

Thus, higher cropping intensity means that a higher portion of the net area is being cropped more than once during one agricultural year. This also implies higher productivity per unit of arable land during one agricultural year. The cropping intensity shows great spatial variation in India, with higher levels in northern plains. Lower levels are found in dry, rain-fed regions of Rajasthan, Gujarat, Maharashtra and Karnataka.

Cropping intensity index (CII)

Cropping intensity index is a measure of farmer's actual land use in area and time relationships for each crop or group of crops compared to the total available land area and time, including the land available for production.

$$CII = \frac{\sum_{i=1}^{Nc} aiti}{A0T \sum_{i=1}^{M} AjTj}$$

where, Nc = total number of crops grown by a farmer during the time period T

ai = Area occupied by the ith crop

ti = Duration of the ith crop

A0 = Farmer's total cultivated land area available for use during the entire time period T

M = Total number of fields temporarily available to the farmer for cropping during time period T

Aj = Land area of jth field;

Tj = Time period that j^{th} field occupies

Specific crop intensity index (SCII)

SCII is a derivative of cropping intensity index and determines the amount of area-time devoted to each crop or group of crops compared to the total time available to the farmer for crop production under study.

$$SCII = \frac{\sum_{k=1}^{Nk} ak\ tk}{A0T + \sum_{j=1}^{M} AjTj}$$

where, Nk = Total number of crops within a specific designation such as vegetable crops or field crops grown by the farmer during the time period T

ak = Area occupied by the kth crop

tk = Duration of the kth crop

A0 = Farmer's total cultivated land area available for use during the entire time period T

M = Total number of fields temporarily available to the farmer for cropping during time period T

Using this formula vegetable intensity index, field crop yield index, rice or wheat crop yield index, etc. can be calculated.

Relative cropping intensity index (RCII)

It is the modification of CII and determines the amount of area and time allotted to one crop or group of crops related to area-time actually used in the production of all crops.

$$RCII = \frac{\sum_{k=1}^{Nk} ak\ tk}{A0T + \sum_{t=1}^{n} aiti}$$

These indices (SCII and RCII) can be used for classifying farmers, based on the crops grown in how much area and for what time period it occupies the field. These indices help to measure shifts of various crops among farm of different sizes and determining whether the consistent types of cropping pattern occur within various farm size strata. These indices also help to know how intensively cultivated land area has been utilised. But none of these indices takes productivity into account and cannot be used for comparing different cropping systems and evaluating their efficiency in utilisation of the resources other than the land.

Rotational intensity (RI)

This is calculated by counting the number of crops in a rotation and multiplied by 100 and then divided by duration of the rotation.

$$RI = \frac{\text{Number of crops grown in a rotation}}{\text{Duration of rotation}} \times 100$$

4.2.2 Assessment of Yield Advantage

Crop equivalent yield (CEY)

Since several crops are grown in intercropping or sequential cropping systems, it is not logical to compare the total yield of different crops in one system with the other. Thus yields from all the crops grown mixed or intercropped or sequentially cropped are converted to the yield of main crop of the system based on price of the produce. The crop equivalent yield is estimated as follows.

$$CEY = \Sigma_{i=1}^{n}(Yi, ei)$$

where, Yi = yield of the ith component and

ei = equivalent factor of ith component which can be calculated as $\frac{Pi}{Pm}$, where Pi is the price of a unit weight of ith crop and Pm is the price of a unit weight of main crop.

This type of comparison is valid only while considering the gross returns. However, it does not indicate the net gain obtained from a cropping system. This also does not explain the land use pattern of the cropping systems.

Relative yield total (RYT)

It is assumed that in inter-genotypic competition the biomass yield of each component is strictly proportional to the share of environmental resources it can acquire. Based on this assumption in a 1:1 ratio, mixtures are competing for same supplies of environmental resources, the proportional increase of per plant yield of one component will tend to equal the proportional decrease of per plant yield of the other. This implies that the mean of plant relative yield (ratio of the per plant yields in a mixture) to that in monoculture will have a value close to unity. However, the use of RYT is restricted to replacement series of intercropping only.

$$RYT = \frac{1}{2}\left(\frac{Yij}{Yii} + \frac{Yji}{Yjj}\right)$$

where, Yij and Yji are yields of i and j species in a mixture respectively and Yii and Jjj are their respective monoculture yield in 1:1 ratio.

Land equivalent ratio (LER)

The concept of land equivalent ratio is the most widely accepted index for evaluating the effectiveness of all forms of mixed cropping. LER is the relative land area under sole crop required to produce the same yield as obtained under a mixed or an intercropping system at the same management level. It is similar to the concept RYT. However, unlike RYT, LER can be used for any intercropping treatments and not just for replacement series. It is calculated as sum total of the ratios of yield of each component crop in an intercropping or mixed cropping system to its corresponding yield when grown as a sole crop.

$$LER = La + Lb = \frac{Ya}{Sa} + \frac{Yb}{Sb}$$

where, La and Lb are the LERs for individual crops in the mixture, Ya and Yb are the individual crop yields in an intercropping situation, Sa and Sb are the yields of species a and b as sole crops.

LER provides a standardising basis so that crops can be added to form combined yield. It can also be used to assess the competitive abilities of the component species of any intercropping situation. The LER has the great merit that it gives an accurate assessment of the greater biological efficiency of the intercropping situation. When LER = 1, there is no advantage to intercropping over sole cropping, when LER < 1, more land is needed to produce a given yield by each component as an intercrop. Intercropping is beneficial only when LER > 1.

Relative crowding coefficient (RCC)

Relative crowding coefficient is a measure of the relative dominance of one component crop over the other in an intercropping or mixed cropping system. The coefficient (K) is determined separately for each component crop.

For crop 'a' in association with crop 'b', the relative crowding coefficient is

$$Kab = \frac{Yab}{Yaa - Yab} \times \frac{Zba}{Zab}$$

where, Y_{ab} is the yield of species 'a' in association with 'b', Y_{aa} is the pure stand yield of species 'a', Z_{ab} and Z_{ba} represent the sown proportion of 'a' (in association with 'b') and 'b' (in association with 'a'), respectively.

Similarly, for crop 'b' in association with crop 'a', the relative crowding coefficient is

$$Kba = \frac{Yba}{Ybb - Yba} \times \frac{Zab}{Zba}$$

where, Y_{ba} is the yield of species 'b' in association with 'a', Y_{bb} is the pure stand yield of species 'b'.

The product of the two coefficients, K_{ab} and K_{ba} designated as K, denotes yield advantage if $K > 1$, no advantage if $K = 1$ and disadvantage if $K < 1$. With the index more than a unity will mean that the given crop is more competitive than the associate crop and vice-versa. Major limitation of RCC is that the comparison of the coefficients for each crop cannot give a quantitative measure of the intercrop competition, but can only indicate that a given crop is more or less competitive.

Aggressivity

Aggressivity gives a simple measure of how much the relative yield increase in species 'a' is greater than that for species 'b' in an intercropping system and can be expressed as Aab.

$$Aab = \frac{Yab}{YaaZab} - \frac{Yba}{YbbZba}$$

where, Yaa and Ybb are pure stand yields of species 'a' and 'b' respectively; Yab and Yba are mixture yields of species 'a' in combination of species 'b' and species 'b' in combination of species 'a' respectively; Zab and Zba are sown proportions of species 'a' in mixture with 'b' and species 'b' in mixture with 'a' respectively.

Aggressivity reveals the simple difference between the extent to which crop 'a' and 'b' vary from their respective expected yields. However, since this is a simple difference, the interpretation of intercropping situations may become difficult if the values are identical in different situations.

Competitive ratio (CR)

Considering the drawback of aggressivity values, the use of competitive ratio has been proposed. Here instead of taking the difference of two terms in aggressivity, the ratio of these terms are taken and designated as competitive ratio (CR).

$$CRa = \frac{Yab}{YaaZab} \div \frac{Yba}{YbbZba}$$

This can also be written as

$$CRa = (\frac{Yab}{Yaa}) \div (\frac{Yba}{Ybb}) \times (\frac{Zba}{Zab}) \text{ or } (\frac{La}{Lb}) \times (\frac{Zba}{Zab})$$

In the above equations, all their abbreviations are the same as in case of aggressivity. However, La and Lb are the LER for crops 'a' and 'b' respectively.

Competition index (CI)

Competition index is the product of two equivalent factors of component crops. Equivalent factor for component 'a' means the number of plants of component 'a' which is equally competitive to one plant of component 'b'. If the equivalence factor is < 1 for a component, the component is more competitive than the other. If the index is < 1, it will be an advantage.

$$\text{Equivalence factor for 'a'} = \frac{\text{yield of 'b'/ plant}}{\text{yield of 'a'/ plant}}$$

$$\text{Equivalence factor for 'b'} = \frac{\text{yield of 'a'/ plant}}{\text{yield of 'b'/ plant}}$$

CI = Equivalence factor for 'a' x Equivalence factor for 'b'

Competition coefficient (CC)

Competition coefficient is the ratio of the relative crowding coefficient (RCC) of any species in the mixture.

$$\text{CC} = \frac{\text{RCC of a given species}}{\text{Total RCC of all crops in mixture}}$$

It is used to find out the relative crowding from which maximum yield can be obtained without any adverse effect on any of the species.

Area time equivalency ratio (ATER)

It is a modification of LER and takes in account the duration of crop and permits an evaluation of crops on yield per day basis. ATER is the ratio of number of hectare-days required in monoculture to the number of hectare-days used in intercropping to produce identical quantities of each of the component crops.

$$\text{ATER} = \frac{\text{La} \times \text{Da} + \text{La} \times \text{Db}}{\text{T}}$$

where, La and Lb are relative yields or partial LERs of component crops 'a' and 'b', Da and Db are duration of crops 'a' and 'b' and T is the total duration of the intercropping system.

ATER > 1 indicates a more efficient use of area-time, e.g. hectare-days by the intercrop.

Relative productivity efficiency

It refers to the capacity of the diversified system for production in relation to the existing system and expressed in percentage.

Relative productivity efficiency =

$$\frac{\text{Total yield of diversified cropping system - total yield of existing cropping system}}{\text{Total yield of existing cropping system}} \times 100$$

Positive figure show the superiority of the new system over the existing one in percentage and considered desirable. Any positive figure of more than 20% is considered worth recommending for extension use.

Sustainable yield index (SYI)

This index is used to assess the sustainability of a cropping system or management practice which can be considered for wider adoption. It identifies the practices systems giving the maximum sustainable yield.

$$SYI = \frac{Y - Sd}{Y\max}$$

where, $\dot{Y}$ is the estimated average yield of different practices or systems over years, Sd is the estimated standard deviation and Ymax is the observed maximum yield among all the systems over years.

Less fluctuation in yield over the years reflects positive side of stability. The stability index ranges from 0 to 1. While calculating SYI, the negative value of (Y – Sd) is taken as zero since yield is always a positive quantity. Sd is used to quantify the risk associated with the average performance of a system. When Sd = 0 and Y = Ymax, SYI is 1. This is an ideal system or practice which gives consistently maximum yield in all the years. But invariably in biological systems Sd is always greater than zero due to variations in yield over years because of climatic factors. The value of SYI nearing to unity shows higher stability reflecting that the system is more sustainable.

4.2.3 Assessment of Economic Advantage

Economic viability of a cropping system can be assessed by converting inputs and products into monetary values. However, the major weakness of this monetisation is that costs and prices of inputs and products rarely remain static and fluctuates from season to season and place to place. Hence, a profitable cropping system in one year may not be so in the next year. For evaluation of cropping systems several indices are being used.

Gross return

This is total income from the farm, by virtue of sales of entire farm produce. The main drawback in this is that the cost involved in production is not considered.

Cost of cultivation

This is a supplementary index to indicate the amount of capital resources needed to adopt a particular cropping system. The resource poor farmer will lower cultivation cost and cropping system with low profit over the system yielding high profit with high investment in many cases.

Net return

It is the income obtained after deducting cost of cultivation from the gross return. The monetary difference of input and output value is more meaningful than the gross return. But variable cost like rent of land, interest on capital, land revenue, etc. are generally excluded while computing net return.

Benefit cost ratio

This index provides an estimate of the benefit a farmer derives for the expenditure incurred in adopting a particular cropping system. This measure shows the return per each rupee investment. Wider the ratio more is the benefit.

$$\text{B: C ratio} = \frac{\text{Gross return}}{\text{Cost of cultivation}} \text{ (From economist point of view)}$$

$$\text{B: C ratio} = \frac{\text{Net return}}{\text{Cost of cultivation}} \text{ (From agronomist point of view)}$$

Relative economic efficiency (REE)

It refers to the capacity of the diversified cropping system for realising net profit in relation to the existing system and expressed in percentage. Positive figure shows that the new system is more profitable over the existing one and considered desirable.

Relative ecnomic efficiency =

$$\frac{\text{Net return of diversified cropping system - net return of existing cropping system}}{\text{Net return of existing cropping system}} \times 100$$

Income equivalent ratio (IER)

The ratio of the area needed under sole cropping to produce the same gross income as one hectare of intercropping at the same management level. IER is the conversion of LER into economic terms.

$$IER = \frac{Iab}{Iaa} + \frac{Iba}{Ibb}$$

where Iab and Iba are the gross income of component 'a' and 'b' in an intercropping situation, Iaa and Ibb are the gross income of component 'a' and 'b' as sole crops.

Relative net return index (RNRI)

RNRI is the ratio between the monetary value of base and intercrop plus differential cost of cultivation of intercropping and sole cropping of base crop and monetary value of base crop as sole crop.

$$RNRI = \frac{Pi\,Yi + Pj\,Yj \pm Dij}{Pi\,Yii}$$

where, Yi and Yj are the yields of ith base crop/ha and j^{th} intercrop/ha, respectively in ij^{th} crop combination; Pi and Pj are the prices of ith base crop and j^{th} intercrop respectively; Y^{ii} is the yield of ith sole crop/ha and Dij is the differential cost of cultivation of ij^{th} crop combination in comparison to i^{th} sole crop.

This index appears to be more appropriate for comparisons in intercropping situations since it takes into account the cost of cultivation and thus, is said to be based on net returns, one of the important requirements for comparisons.

Sustainable value index (SVI)

This index is used to assess the sustainability of a cropping system on monetary terms. In cropping systems where more than one crop is involved, the economic assessment is ideal one than the biological assessment. To work out the SVI, the monetary values of the economic produce are used instead of yield values. It is worked out by the same way as SYI.

$$SVI = \frac{Y - Sd}{Ymax}$$

where, Y is the estimated average net profit of a system over years, Sd is the estimated standard deviation and Ymax is the observed maximum profit among all the systems over years.

5

Sustainable Agriculture

The Green Revolution started in the early 1960s that led to the attainment of self-sufficiency in food grain production. This has been considered as the greatest agricultural transformation in the history of humankind. The Green Revolution contributed to widespread poverty reduction, averted hunger for millions of people, and avoided the conversion of thousands of hectares of land into agricultural cultivation. At the same time, it also spurred its share of unintended negative consequences, often not because of the technology itself but rather, because of the policies that were used to promote rapid intensification of agricultural systems and increase food supplies. Some areas were left behind, and even where it successfully increased agricultural productivity, the Green Revolution was not always the panacea for solving the myriad of poverty, food security, and nutrition problems facing poor societies (Pingali, 2012). The benefits of Green Revolution have been poorly distributed and thus, hunger still persists in many parts of the world.

5.1 CONCEPT OF SUSTAINABLE AGRICULTURE

The modern high-input agriculture has developed several adverse effects, more particularly on ecological aspects. Some of these adverse effects are listed below.

1. Overuse of natural resources, causing depletion of groundwater, and loss of forests, wild habitats, and of their capacity to absorb water, causing waterlogging and increased salinity
2. Contamination of the atmosphere by ammonia, nitrous oxide, methane and the products of burning, which play a role in ozone depletion, global warming and atmospheric pollution
3. Contamination of food and fodder by residues of pesticides, nitrates and antibiotics
4. Contamination of water by pesticides, nitrates and agrochemicals causing human and animal health hazards, harm to wildlife, disruption of ecosystems
5. Build-up of resistance to pesticides in pests and diseases including herbicide resistance in weeds

6. Erosion of genetic diversity by focusing on modern varieties, and causing the displacement of traditional varieties and breeds
7. New health hazards for workers in the agrochemical and food processing industries

Added to the above adverse effects, the increasing human as well as livestock population imposed intense pressure on available natural resources. Accordingly, a challenge emerged that required a new vision, holistic approaches for ecosystem management and renewed partnership between science and society. In December 1983, the United Nations General Assembly established the World Commission on Environment and Development under the leadership of former Norwegian prime minister Harlem Brundtland. In 1987, the 'Brundtland Commission' released its final report, *Our Common Future*. It was then realised that after decades of effort to raise living standards through industrialisation, many countries were still dealing with extreme poverty. It seemed that economic development at the cost of ecological health and social equity did not lead to long-lasting prosperity. It was also realised that our generation has too often been willing to use the resources of the future to meet our own short-term goals. It is a debt we can never repay. If we fail to change our ways, our future generations would suffer more than we, and they would be denied their fundamental right to a healthy, productive, and life-enhancing environment. Because resources on the earth are limited, it was clear that the world needed to find a way to harmonise ecology with prosperity. The document famously defines sustainable development as development that meets the needs of the present without compromising the ability of future generations to meet their own needs (WCED, 1987). The Commission successfully unified environmentalism with social and economic concerns on the world's development agenda.

Almost at the same time the realisation of prime importance of staple food production for achieving food security for future generations has brought the concept of 'Sustainable Agriculture' to the forefront and began to take shape in the following three points.

1. The interrelatedness of all the farming systems including the farmer and the family
2. The importance of many biological balances in the system
3. The needs to maximise desired biological relationships in the system and minimise the use of materials and practices that disrupt these relations

In fact, the idea of sustainable agriculture has been around a long time. Since the very first crop was sown and animal was penned, farmers have tried to ensure that their land produces a similar or increasing yield of products year after year. The recent attempts to popularise the concept of sustainable agriculture build on this tradition. Sustainable agriculture has evolved from three perspectives; as a system of production to achieve food self-reliance; as a concept of stewardship; and as a vehicle for sustaining rural communities (MacRae, 1990). A sustainable agriculture is a system of agriculture that will last. It is an agriculture that maintains its productivity over the long run. Sustainable agriculture is both a philosophy and a system of farming. It has its roots in a set of values that reflects an awareness of both ecological and social realities. It involves design and management procedures that work with natural processes to conserve all resources, minimise waste and environmental damage, while maintaining or improving farm profitability. Working with natural soil processes is of particular importance. Sustainable agriculture systems are designed to take maximum advantage of existing soil nutrient and water cycles, energy flows, and soil organisms for food production. Such systems aim to produce food that is nutritious, without being contaminated with products that might harm human health.

Table 5.1. Difference between sustainable and modern agriculture

Particulars	Sustainable agriculture	Traditional agriculture
Plant nutrients	Farm yard manures, compost, vermicompost, green manure, bio-fertiliser, and crop rotation	Chemical fertilisers
Pest control	Cultural methods, crop rotation and biological methods	Chemical pesticides
Inputs	High diversity, renewable and biodegradable inputs	High productivity and low diversity
Ecology	Stable ecology	Fragile ecology
Use of resources	Extraction of natural resources does not exceed the rate of regeneration	Extraction exceeds regeneration, over-exploitation of natural resources
Quality of produce	Safe for human and animal consumption	Presence of toxic residue, unsafe for consumption

Source: https://agriinfo.in/classifications-of-cropping-system-645/

The Food and Agriculture Organisation of the United Nations has established five basic principles for the global agricultural sector to become increasingly productive and sustainable (FAO, 2014). The intention of these five principles is to build a production system that works in favour of the ecosystem, satisfying the human needs. For that, the farmers should work in the economic development of this sector and, so, they can be benefited from a fair, equitable and efficient environment.

Principle 1: Improving efficiency in the use of resources is crucial to sustainable agriculture

Principle 2: Sustainability requires direct action to conserve, protect and enhance natural resources

Principle 3: Agriculture that fails to protect and improve rural livelihoods, equity and social well-being is unsustainable

Principle 4: Enhanced resilience of people, communities and ecosystems is key to sustainable agriculture

Principle 5: Sustainable food and agriculture requires responsible and effective governance mechanisms

These five principles are complementary; Principle 1 and Principle 2 directly support the natural system, while Principle 3 directly supports the human system; Principles 4 and Principle 5 underpin both the natural and human systems. For application of the five principles, a range of actions should be taken to enhance sectoral as well as cross-sectoral productivity and sustainability. To be sustainable and productive, agriculture will need to adopt a vision that maximises synergies, mitigates negative externalities and minimises harmful competition between its sectors.

5.2 DEFINITIONS OF SUSTAINABLE AGRICULTURE

Sustainable agriculture refers to an agricultural production and distribution system that

1. achieves the integration of natural biological cycles and controls
2. protects and renews soil fertility and the natural resource base
3. reduces the use of nonrenewable resources and purchased (external or off-farm) production inputs
4. optimises the management and use of on-farm inputs
5. provides adequate and dependable farm income
6. promotes opportunity in family farming and farm communities
7. minimises adverse impacts on health, safety, wildlife, water quality and the environment

There is no generally accepted definition of sustainable agriculture. However, as Swindale (1988) explained, sustainability conveys the idea of a balance between human needs and environmental concerns. A common theme among definitions is that sustainable agricultural systems remain productive over time

(Senanayake, 1991). The enhancement of the environmental quality and careful use of the resource base on which agriculture depends is viewed as a requisite to sustained agricultural productivity (ASA, 1989). Among the topics considered under sustainable agriculture are resource management issues dealing with soils, land, natural resources and watersheds; and environmental problems such as desertification, soil degradation, etc. (Davis and Schirmer, 1987).

Sustainable agriculture is also defined as successful management of resources for agriculture to satisfy changing human needs while maintaining or enhancing the natural resource-base and avoiding environmental degradation (CGIAR-TAC, 1988).

A sustainable agriculture is a system of agriculture that is committed to maintain and preserve the agriculture base of soil, water, and atmosphere ensuring future generations the capacity to feed them with an adequate supply of safe and wholesome food. A sustainable agriculture system is one that can indefinitely meet demands for food and fibre at socially acceptable, economic and environment cost (Crosson, 1992).

FAO (1988) defines sustainable agriculture as an agricultural practice that conserves land, water, and plant and animal genetic resources, and is environmentally non-degrading, technically appropriate, economically viable and socially acceptable. The management and conservation of the natural resource base and the orientation of technological change should be in such a manner as to ensure the attainment of continued satisfaction of human needs for present and future generations.

Lal (1991) defined sustainable agriculture as an increasing trend in production over time per unit consumption of the non-renewable or the limiting resource, or per unit degradation of soil and environmental characteristics.

Board for International Food and Agricultural Development Task Force defined sustainable agriculture as the successful management of resources for agriculture to satisfy changing human needs, while maintaining or enhancing the natural resource base and avoiding environmental degradation (BIFAD, 1988).

Okigbo (1991) defined sustainable agriculture as a system in which the farmer continuously increases productivity at levels that are economically viable, ecologically sound, and culturally acceptable, through the efficient management of resources and orchestration of inputs in numbers, quantities, sequences and timing with minimum damage to the environment and danger to human life.

These definitions can be supplemented by some fundamental principles of sustainable agriculture.

1. Farm productivity is enhanced over the long term.
2. Adverse impacts on the natural resource base and associated ecosystems are ameliorated, minimised or avoided.
3. Residues resulting from the use of chemicals in agriculture are minimised.
4. Net social benefit (in both monetary and non-monetary terms) from agriculture is maximised.
5. Farming systems are sufficiently flexible to manage risks associated with the vagaries of climate and markets.

Sivakumar et al. (2000) summarised the definitions of 'sustainable agriculture' given by several authors and organisations and concluded that 'natural resource use' is a keyword common to almost all of them.

5.3 OBJECTIVES AND GOALS OF SUSTAINABLE AGRICULTURE

Objectives

1. To make best use of the resources available.
2. To minimise use of non-renewable resources.
3. To protect the health and safety of farm workers, local communities and society.
4. To protect and enhance the environment and natural resources.
5. To protect the economic viability of farming operations.
6. To provide sufficient financial reward to the farmer to enable continued production and contribute to the well-being of the community.
7. To produce sufficient high-quality and safe food.
8. To use the available technology, knowledge and skills to suit local conditions.

Goals

Pretty (1996) identifies the following goals of sustainable agriculture.

1. Thorough incorporation of natural processes such as nutrient cycling, nitrogen fixation and pest-predator relationships into agricultural production processes.
2. Reduction in the use of those off-farm, external and nonrenewable inputs with the greatest potential to damage the environment or harm the health of farmers and consumers, and more targeted use of the remaining inputs used with a view to minimising variable costs.

3. Equitable access to predictive resources and opportunities, and progress towards more socially just forms of agriculture.
4. More productive use of local knowledge and practices, including innovation in approaches not yet fully understood or widely adopted.
5. Increase in self-reliance among farmers and rural people.
6. Improvement in the match between cropping patterns and the productive potential and environmental constraints of climate and landscape to ensure long-term sustainability of current production levels.
7. Profitable and efficient production with an emphasis on integrated farm management and the conservation of soil, water, energy and biological resources.
8. Full participation of farmers and rural people in all processes of problem analysis and technology development, adoption and extension.
9. Greater productive use of the biological and genetic potential of plant and animal species.

5.4 THREE PILLARS OF SUSTAINABILITY

The adjectives biological, ecological, alternative, regenerative, and low-input are commonly used to refer to seemingly similar concepts of agricultural systems. The term sustainable agriculture broadly includes all of these concepts to some degree, and also addresses a specific set of criteria. Thus, three broad areas of concern seem to underlie the concept of sustainable agriculture (Weil, 1990).

1. Economic concerns over economic justice, the survival of owner operated farms, and the long-term profitability of agriculture.
2. Environmental concerns over adverse impacts of agriculture on land, water, and wildlife resources.
3. Public welfare concerns over food quality and human exposure to toxic chemicals.

Basically, sustainable agriculture is a philosophy based on human goals and on understanding the long-term impact of our activities on the environment and other species (Francis, 1990). Broad concepts in sustainable agriculture encompass ecological, economic, and social parameters. Therefore, sustainability is a holistic approach that considers ecological, social and economic dimensions, recognising that all must be considered together to find lasting prosperity (Weil, 1990).

Environmental sustainability

Ecological integrity is maintained, all of earth's environmental systems are kept in balance while natural resources within them are consumed by humans at a rate where they are able to replenish themselves.

Economic sustainability

Human communities across the globe are able to maintain their independence and have access to the resources that they require, financial and other, to meet their needs. Economic systems are intact and activities are available to everyone, such as secure sources of livelihood.

Social sustainability

Universal human rights and basic necessities are attainable by all people, who have access to enough resources in order to keep their families and communities healthy and secure. Healthy communities have just leaders who ensure personal, labour and cultural rights are respected and all people are protected from discrimination.

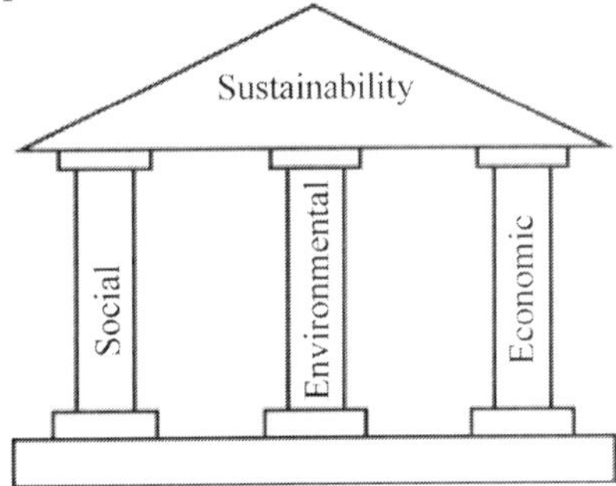

Fig. 5.1. Three pillars of sustainability

Source: https://www.calsense.com/three-pillars-of-sustainability/

5.5 ELEMENTS OF SUSTAINABLE AGRICULTURE

Weil (1990) proposed three sets of criteria by which the direction, toward greater or less agricultural sustainability, may be judged. To be judged sustainable, changes to the current system should be an improvement over present practice and policy by all three sets of criteria. An agricultural programme, policy, or practice contributes to agricultural sustainability if it

1. Enhances or maintains the number, quality, and long-term economic viability of farming and other agricultural business opportunities in a community or region;

2. enhances, rather than diminishes, the integrity, diversity, and long-term productivity of both the managed agricultural ecosystem and the surrounding natural ecosystems; and
3. enhances, rather than threatens, the health, safety, and aesthetic satisfaction of agricultural producers and consumers alike.

Using these criteria, certain inputs or practices may be judged 'sustainable' in one situation but not in another. There are several means that are generally, but not always, relevant to achieving greater agricultural sustainability. There are many ways to improve the sustainability of a given farming system, and these vary from region to region, However, there are some common sets of practices among farmers trying to take a more sustainable approach, in part through greater use of on-farm or local resources each contributing in some way to long-term profitability, environmental stewardship and rural quality of life. Weil (1990) has listed the following elements that increase sustainability in agriculture.

Biological diversity

Species diversity is usually a sign of ecological stability. A diversity of crop and animal enterprises on a farm also often lends economic stability. Growing a number of crops on a farm can help reduce risks from extremes in weather, market conditions or crop pests. Increased diversity in crops and other plants, such as trees, shrubs and other woody perennials, also can contribute to soil conservation, wildlife habitat and increased populations of beneficial insects. Agroforestry systems such as agrisilviculture, silivipastoral, agrisilvihorticultural, hortisilvipastoral, alley cropping, etc. may be practised to achieve both ecological and economic stability.

Crop rotations

Along with increasing diversity, crop rotations are known to reduce soil erosion, ameliorate pest problems, improve soil fertility when legumes are included in the rotation, and also produce an additional, unaccounted for 10 to 15% increase in crop yields.

Animal integration

Plant and animal production are so complimentary that integration of animals into farming systems often seems essential to developing a sustainable system. Concentration of livestock production separate from crop production makes it difficult to justify the kinds of rotations that enhance sustainability. It also makes manure a serious waste disposal problem instead of a valuable on-farm resource. New management-intensive grazing systems may be introduced to take animals out barn into the pasture to provide high-quality forage and reduced feed cost. Rotational grazing improves the sustainability in forage production in the pasture.

Cover crops

Growing plants such as sunhemp, horsegram, cowpea in the off-season after harvesting a grain or vegetable crop can provide several benefits, including weed suppression, erosion control, and improved soil nutrients and soil quality.

Soil conservation and soil health

Many soil conservation practices, including contour cultivation, contour bunding, graded bunding, vegetative barriers, strip cropping, cover cropping, reduced tillage, etc. help prevent loss of soil due to wind and water erosion. The soil management practices should be designed to enhance the viability and diversity of soil organisms (microbes, arthropods, earthworms, etc.) and biological functions (organic matter and nutrient cycling, etc.)

Nutrient management

Proper management of nitrogen and other plant nutrients can improve the soil and protect environment. Increased use of farm nutrient sources such as manure and leguminous cover crops also reduces purchased fertiliser costs.

Integrated pest management

Integrated pest management is a sustainable approach to managing pests by combining biological, cultural, physical and chemical tools in way that minimises economic, health and environmental risks.

Knowledge-based farming

Sustainable agriculture will probably depend increasingly on sophisticated information on weather prediction, pest life cycles, soil processes, markets, and the like, melded into a holistic understanding of the agroecosystem and the agricultural community. An important aspect of this arena is the ability of the individual farmer to assimilate and, in many cases, to generate the information and insights needed to allow him or her to adapt to changing circumstances. Sustainability implies a dynamic, self-adjusting system.

Human scale farm size

There is no simple answer to how big is too big for sustainability, but the trend of the past few decades toward ever larger farms, operated by ever fewer farmers who are ever more remote from their fields is one that does not stand up well to the criteria of sustainability, both in terms of substituting management for non-renewable resources, and in terms of sustaining the numbers of economic opportunities in farming.

Minimal dependence on non-renewable resources

The basic resources of agriculture are sunlight, air (carbon and oxygen), water, soil (including nutrient elements therein), germplasm (of animals, plants and microbes), and farmers (with their skills, knowledge and labour). These resources are renewable and can sustain agriculture in perpetuity, if managed well. Agriculture cannot be sustained in the long run if it depends on the one-way, dead-end street of non-renewable resources use; e.g. burning fossil fuels, dispersing pesticides made from fossil fuels, or mining nutrients and dispersing them without recycling. Soil can only be considered renewable if not allowed to erode away faster than it forms.

Marketing

Farmers realise that improved marketing is required for their produce to enhance profitability. Without a favourable market the economic aspect of agriculture may not be sustainable. Direct marketing of agricultural product from farmers to consumers should be promoted.

5.6 ADVANTAGES OF SUSTAINABLE AGRICULTURE

Contributes to environmental conservation

Sustainable agriculture helps to replenish the land as well as other natural resources such as water and air. By adopting sustainable practices, farmers will reduce their reliance on nonrenewable energy, reduce chemical use and save scarce resources. This replenishment ensures that these natural resources will be able to sustain life for future generations considering the rising population and demand for food.

Saves energy for future

Modern agriculture is heavily dependent on nonrenewable energy sources, especially petroleum. Sustainable agricultural systems have reduced the need for fossil fuels or nonrenewable energy sources and a substitution of renewable sources to the extent that is economically feasible.

Diversifies crops and farm products

Diversity of crops and farm products promotes recycling of products and by-products and thus, reduces the need for external high cost inputs. Diversity buffers the farm against sudden fall in prices of one or a few farm products and increase in costs of production inputs. It also buffers the farm against unpredictable weather. With sustainable agriculture, it is possible to produce more than one product in small areas at a time with high efficiency.

Increases self-reliance

Sustainable agriculture uses locally available renewable resources, appropriate and affordable technologies and minimises the use of external and purchased inputs, thereby increasing self-sufficiency and self-reliance, and insuring a source of stable income for the farmers. This encourages people to stay on farming.

Public health safety

Sustainable agriculture avoids hazardous pesticides and fertilisers. As a result, farmers are able to produce fruits, vegetables and other crops that are safer for consumers, workers, and surrounding communities. Through careful and proper management of livestock waste, sustainable farmers can protect humans from exposure to pathogens, toxins, and other hazardous pollutants.

Prevents pollution

Sustainable agriculture means that any waste a farm produces remains inside the farm's ecosystem. Organic farm wastes are converted to useful manures through the process of composting and vermicomposting. In this way, the waste cannot cause pollution.

Prevents air pollution

Agricultural activities affect air quality by smoke from agricultural burning; dust from tillage, traffic and harvest; pesticide drift from spraying; and nitrous oxide emissions from the use of nitrogen fertiliser. In sustainable agriculture, there are options to improve air quality by incorporating crop residue into the soil, using appropriate levels of tillage, and planting windbreaks, cover crops or strips of native perennial grasses to reduce dust.

5.7 THREATS TO AGRICULTURAL SUSTAINABILITY

5.7.1 Land Degradation

One of the most critical threats to sustainable agriculture is land degradation. Vast areas of cropland, grassland, woodland and forest in Asia and Pacific have already been lost, and many more are threatened. The land degradation has both on-site and off-site effects. On-site effects include lowering of productive capacity causing either reduced outputs or need for increased inputs. Off-site effects of water erosion occur because of the deterioration in water quality, river sedimentation, biodiversity loss, and natural calamities. The irrigated agriculture, especially through canal systems, has resulted in land degradation at many places due to the twin problems of waterlogging and salinisation. A decline in soil productivity, particularly of organic carbon and nitrogen, deterioration in soil physical characteristics, and decreasing water availability are identified as the causes of this slowdown in productivity.

Much of the land degradation results from unstable use of land, over intensive cultivation and inappropriate management practices. In order to meet their basic food needs, smallholders and the rural poor have been pushed into using ecologically fragile areas, forced to crop intensively on steep slopes that are vulnerable to erosion. Land degradation has also resulted from excessive use of mineral fertilisers and over-intensive livestock-keeping. Pesticides and heavy metals diminish soil respiration, microbial activity, soil chemical action, restrain significant soil procedures like ammonification and nitrification, decrease earthworm dominance, and smother algal activity. These possibly unsafe substances may accumulate in soil and cause long-term consequences for yield and quality, and may harm soil microflora.

Lal (1991) estimated that the per capita arable land will progressively decline from about 0.3 ha in 1990 to 0.1 ha by 2050, 0.14 ha by 2100. The availability of per capita net sown area in India has reduced from 0.33 ha in 1951 to 0.14 ha in 2001 and is expected to decline further to 0.09 ha by 2050. A minimum economic holding size of 2 ha of un-irrigated land and 1ha of irrigated land has been suggested in India for sustaining a family of 5 or 6 persons (Ghosh et al., 2020).

5.7.2 Water and its Availability

Rice requires two or three times more water for cultivation than other cereals. Rice being the staple food across Asia and the Pacific region, agriculture is the principal user of water, consuming around 70% of total withdrawals. It is estimated that by 2050, Asia and Pacific would need an additional 2.4 billion cubic meters of fresh water/day. World population is expected to increase to over 10 billion by 2050, and this population will need food and fibre to meet its basic needs. Combined with the increased consumption of calories and more complex foods, which accompanies income growth in the developing world, it is estimated that agricultural production will need to expand by approximately 70% by 2050 (World Bank, 2020). While 2 litres of water are often sufficient for daily drinking, it takes about 3,000 litres to produce the daily food needs of one person.

In India, the utilisable surface-water resource is estimated to reduce by 7% due to deforestation and soil erosion, while the loss due to water pollution is put at

20%. Some of the overuse of water in agriculture reflects government policy. Many governments have subsidised the construction of inefficient irrigation systems, along with free electricity supply. This encourages farmers to withdraw too much water from rivers, over-pump groundwater and generally waste freshwater resources. In addition, intensive agriculture and industrial effluents are creating high levels of pollution. Hence, there is an urgent need to preserve and maintain the water quality of surface and groundwater resources for production and other purposes. It is estimated that by 2050 about 22% of the geographical area and 17% of the population would be under absolute water scarcity. Paradigm changes would be required in water resources development and management to avert such worst scenario. The problem of the falling groundwater table in central Punjab, where rice is a predominant crop, is because of the overdraft of water. A holistic strategy is required to overcome the water crisis, which includes crop diversification, delayed transplanting of rice, and adoption of water-saving agronomic practices, etc. (Ghosh et al., 2020)

5.7.3 Deforestation

Forests provide critical ecosystem services to the agricultural sector, including pollination and watershed protection, and support to fisheries. Millions of poor people and small-scale enterprises depend on forests for food, fibre, fodder and other materials, but are finding this increasingly hard as the natural forests are shrinking. Human activity has always modified the natural ecosystems in such a way that environment becomes more favourable for non-agricultural activities. Apart from diversion of forest area to non-forest activities like industries, infrastructure development and human habitation, unrestricted exploitation of timber as well as other wood products for commercial purposes, and slash and burn method of cultivation are major causes of forest degradation (Ghosh et al., 2020). Due to deforestation, watersheds that once provided drinking water and irrigation water have now become unreliable due to extreme fluctuations in water flow. Another serious consequence of deforestation is the loss of biodiversity, i.e. the extinction of thousands of species and varieties of plants and animals. Global warming is another consequence of deforestation.

The Asia-Pacific region has around half the world's total area of mangroves. These are under severe strain as a result of the extraction of timber and coastal development, including for the production of environmentally damaging, export-oriented cultivation of shrimps. According to the India State of Forest Report, 2017 the total forest and tree cover is spread across 24.39% of the geographical area of the country as against the target of 33%.

5.7.4 Depletion of Soil Organic Carbon

The carbon stores in arid and semiarid lands show a high temporal and spatial variability, some parts acting as C sources and others as C sinks. A decline in soil organic carbon (SOC) content is a common phenomenon when land use changes from natural vegetation to cropping, reasons being reduction in C inputs, increased rate of decomposition due to mechanical disturbance of the soil, higher soil temperatures due to exposure of the soil surface, more frequent wetting and drying cycles and increased loss of surface soil rich in organic matter through erosion (Ghosh et al., 2020).

Soil organic carbon content in most Indian soils is low but it is also dynamic in nature. Farming practices affect both quantity and quality of organic matter. Lands deteriorated due to depletion of SOC are widely distributed across the country, ranging from cultivated areas of subtropical belt to the areas under shifting cultivation. This problem is acute in high-intensive cultivated areas of rice-wheat cropping system in Indo-Gangetic plains. In the major rice-wheat regions of north-western India, SOC has decreased from 0.5% in 1960s to 0.2% in 2010s (Bhattacharyya et al., 2015). The removal or *in situ* burning of crop residues, no or least addition of organic manures, and intensive cultivation are major reasons for depletion of SOC.

5.7.5 Loss of Biodiversity

Agrobiodiversity is the backbone of a nation's food security and the basis of economic development as a whole. However, the rapid and large-scale global extinction of species also caused loss of biodiversity. In the 20th century, it happened thousand times higher than the average rate during the preceding 65 million years. Overexploitation, habitat destruction, pollution and species extinction are major causes of biodiversity loss. Other factors include fires, which adversely affect regeneration in some cases. Agricultural biodiversity is also at risk from climate change. It is reported that genetically modified crops threaten human health and the environment and will allow large corporations to tighten their grip over agricultural production by narrowing down the wide range of varieties now available just to a few.

Over the years biodiversity in India is under pressure due to the massive commercialisation of agriculture leading to almost extinction of traditional farming systems. Indian farmers grew more than 30,000 different varieties of rice before green revolution. Unfortunately, this enormous diversity has reduced during post-green revolution period. There has been a loss of several thousand rice varieties. Species and communities at particular and possibly critical risk include those with limited climatic ranges, limited dispersal ability and those with specialised habitat requirements (Upadhyay et al., 2008).

5.7.6 Climate Change

Climate change is the most dominant environmental concerns to the agricultural planning. Food security is also being threatened by climate change which will have many complex effects, bringing advantages in some places, but disadvantages in many others. For example, higher concentrations of carbon dioxide in atmosphere may increase photosynthesis in several crops, thereby increasing yields. However, this potential gain could easily be outstripped by the effects of higher temperatures and more variable rainfall. The carbon dioxide has the least global warming potential among major greenhouse gases but due to its much higher concentration in the atmosphere, it is the major contributor towards global warming and climate change. Moreover, this may create significant uncertainties for agricultural production due to longer dry or rainy season, intense heat during summer, more or scanty rains during rainy months, frequent floods, droughts, storms, etc.

The changes in temperature, precipitation, CO_2 concentration, changes in frequency of infestation by pests and diseases caused agriculture vulnerable. In the case of an annual crop, the duration between sowing and harvesting will shorten and the crop may experience terminal heat stress.

The shortening of crop duration may have an adverse effect on productivity. In India, impact of 1-2° C increase in mean air temperature is expected to decrease rice yield by about 0.75 t/ha in efficient zones and 0.06 t/ha in coastal regions; and impact of 0.5° C increase in winter temperature is projected to reduce wheat yields by 0.45 t/ha. Productivity loss of 4–6% in rice, 6% in wheat, 18% in maize, 2.5% in sorghum, 2% in Indian mustard and 2.5% in potato are projected due to climate change (Naresh Kumar et al., 2012). The diversity and dominance of weeds and pests are likely to increase with climate change. Higher temperature restricts the seed germination and also detrimental to the fruit setting in trees (Ghosh et al., 2020).

5.8 INDICATORS OF AGRICULTURAL SUSTAINABILITY

Indicators of agricultural sustainability are a composite set of attributes or measures that embody a particular aspect of agriculture. These are quantified information, which help to explain how things are changing over time. An indicator is a quantitative measure against which some aspect or aspects of policy performance or management strategy can be assessed. Sustainability indicators look at economic, social and environmental information in an integrated manner. Gallopín (1997) identified the following functions of indicators.

1. To assess conditions and changes
2. To compare across place and situations

3. To assess conditions and trends in relation to goals and targets
4. To provide early warning information
5. To anticipate future conditions and trends

5.8.1 Criteria for Selecting Indicators

At least the following three criteria should guide the development of sustainability indicators.

Policy relevance: Indicators should address the issues of primary concern to a country or district and receive the highest priority. In some cases policy makers may already share concern about an aspect of sustainability and be ready to use indicator information for addressing the issue.

Predictability: Indicators should allow a forward-looking perspective that can promote planning and decisions on issues before they become too severe. Anticipatory decision-making is at least as important to sustainable agriculture as is recognition of existing problems.

Measurability: Indicators should allow planners and analysts the means to assess how these were derived, either qualitatively or quantitatively, and decide how these can best be applied in the planning and decision-making process.

Selection of effective indicators is the key to the overall success of any monitoring programme. An indicator must be selected carefully so that it can measure and describe explicitly the condition of sustainability.

The ecological indicators should have the following characteristics (Dale and Beyeler, 2001). The indicators should

1. Be easily measurable,
2. Be sensitive to stresses on the system,
3. Respond to stress in a predictable manner,
4. Be anticipatory, meaning that they signify an impending change in the ecological system,
5. Predict changes that can be averted by management actions,
6. Be integrative, meaning that the full suite of indicators provides a measure of coverage of the key gradients across the ecological systems (such as soils, vegetation types and temperature),
7. Have a known response to natural disturbances, anthropogenic stresses, and changes over time, and
8. Have low variability in response.

Operational indicators for measuring agricultural sustainability may be categorised under three components of sustainability.

Economic

- Crop productivity
- Net farm income
- Benefit-cost ratio of production
- Per capita food grain production

Social

- Food self sufficiency
- Equality in income and food distribution
- Access to resources and support services
- Farmers, knowledge and awareness of resource conservation

Ecological

- Amount of fertilisers / pesticides used per unit of cropped land
- Amount of irrigation water used per unit of cropped land
- Soil nutrient content
- Depth of groundwater table
- Quality of groundwater for irrigation
- Water use efficiency
- Nitrate content of groundwater and crops

Howlett (1996) identified the following purposes of the use of indicators at the farm level.

1. To develop capacity and commitment of farmers towards more sustainable land use, and to allow farmers to evaluate their own practices.
2. For the simple diagnosis of problems and improvements to farming practices, and development of appropriate research and extension activities.
3. To improve the relationship between the researcher, farmer and extension agent, and through this to encourage farmer participation, the incorporation of indigenous knowledge, and ultimately to an increase in the adoption of improved technologies.
4. To assess and monitor the spatial and temporal sustainability of different farming systems, and to use this for the evaluation, prediction, planning and management of these systems by farmer, researcher, extension agent and planners.

Measuring sustainability is most challenging and complex. Rural Industries Research and Development Corporation, Australia indicated general broad measurable components under each hierarchical level of components of sustainability (RIRDC, (1997). Hoang (2013) had analysed productive performance of crop production systems in an integrated analytical framework considering economic, institutional, physical, social and technological factors and indicated that in a dynamic analysis to make efficiency framework to be forward looking, climate change innovations in crop science to be incorporated.

Table 5.2. Sustainability indicators

Hierarchical level	**Sustainability indicators (Economic, social and environmental)**
Cropping system/ farming system	Non-negative trends in 1. Farm productivity 2. Net farm income 3. Total factor productivity 4. Nutrient balance 5. Soil quality 6. Residues in soil, plant, products 7. Farm water use efficiency 8. Farmer skills and education 9. Debt service ratio 10. Health 11. Time spent on other social cultural activities
Agroecosystem (watershed, agroecozone, etc.)	Non-negative trends in 1. Regional production 2. Regional income 3. Regional total factor productivity 4. Regional nutrient balance 5. Income distribution 6. Species diversity 7. Soil loss 8. Surface water quality 9. Groundwater quality 10. Regional social and economic development indicators
Global, national, regional systems	Indefinitely meet the demands at acceptable social, economic and environmental costs.

Source: RIRDC, 1997

Table 5.3. Indicators of key natural resources in rainfed cropping systems

Indicators	Key management aspects
Nutrient balance	Organic matter - rate of change
	Nitrogen cycling - especially when using grain legumes in rotation with cereals
	Monitoring status of phosphorus, sulphur and potassium
	Micronutrients
Erosion	Vegetation cover - includes trees as well as stubble
	Soil surface cover - stubble retained (30% sufficient to prevent wind and water erosion)
	Stream bank
	Sheet and gully erosion
Productivity, yield and quality	Water use efficiency, i.e. actual versus potential (in some areas the potential is much less than the actual) (biomass/grain yield/ net return), recharge (dryland salinity and nutrient leaching)
	Pasture composition - legume and perennial
	Matched animal versus pasture production - appropriate enterprise selection/capability
	Maintenance of genetic base/improvement
Soil structure	Infiltration
	Permeability/water storage
	Stability
	Waterlogging
	Compaction
*p*H	Change
	Toxicity - deficiency
	Indicator plants
Energy efficiency	Energy input vis-a-vis energy output of the whole agricultural system
Biological factors	Soil macro/micro flora and fauna
	Animal health
	Plant health (root growth and other)
	Pests (animals and plants)
Farm management skills	Understanding - a good indicator would be the understanding of the farmers of their own technical system
Rainfall/ precipitation	Performance of rainfall in a year as % of normal and its coefficient of variation
	Distribution of area based on rainfall amount
	Categorisation of the amount of rainfall
	Number of districts having mean annual rainfall of 750-1250 mm and moisture availability period for at least 150 days
	Area affected due to drought

Source: Subba Rao and Mandal, 2007

5.8.2 Indicator Frameworks

Several sets of methodological frameworks or guidelines have been identified for the measurement of sustainability indicators at the farm or community levels. These have all tended to come from an approach focussed on sustainable agriculture and/or sustainable land management. The most widely accepted framework for agricultural sustainability is referred to as Pressure-State-Response (PSR framework). This framework is used by Organisation for Economic Cooperation and Development (OECD), Scientific Committee on Problems of the Environment (SCOPE) and some other organisations working in the field.

Pressure-state-response framework

The Pressure-State-Response framework can facilitate better understanding of actions and activities that are affecting the state of the system, and appropriate response for addressing them both for the agency and stakeholders. The Pressure-State-Response (PSR) framework was conceived by Statistics Canada (Friend and Rapport, 1979), then further developed and adopted internationally in many countries (Waheed et al., 2009).

Pressure refers to the driving forces that create environmental impacts; state refers to the condition(s) that prevail when a pressure exists and response refers to the mitigation action(s) and levers that could be applied to reduce or eliminate the impacts. Thus, a PSR framework states that human activities exert pressure (such as pollution emissions or land use changes) on the environment, which can induce changes in the state of the quality and quantity of the environment (such as changes in ambient pollutant levels, habitat diversity, water flows).

Table 5.4. Examples of PSR framework

Issue	Pressure (driving force)	State (resulting condition)	Response (mitigating action)
Soil erosion	Hillside farming	Declining yield	Terracing, perennial cropping
Water quality	Agro-industrial processing	Fish die-off	Water treatment, technology adjustment
Condition of grassland	Livestock grazing	Soil erosion	Stock rotation, de-stocking, revegetation
Environmental pollution	CO_2 emissions	Rising global temperatures	Introduction of energy taxes

Source: Subba Rao and Mandal, 2007

5.8.3 SDG Indicator 2.4.1

Agriculture plays an essential role in ensuring a better future for all. As a fundamental connection between people and the planet, it can help achieve multiple sustainable development goals (SDGs). Yet, to ensure enough food is produced for a population of nearly 10 billion by 2050 without critically degrading natural resources, a transition to sustainable agricultural systems is needed. FAO has set the following five key principles to achieve the sustainable development goals.

1. Increase productivity, employment and value addition in food systems
2. Protect and enhance natural resources
3. Improve livelihoods and foster inclusive economic growth
4. Enhance the resilience of people, communities and ecosystems
5. Adapt governance to new challenges

In September 2015, the United Nations General Assembly adopted the 2030 Development Agenda and an associated 17 Sustainable Development Goals (SDGs). The resultant SDGs are accompanied by 169 targets under the various goals and a set of 232 indicators to monitor progress toward the SDGs. Responsibility for the development of indicators is given to the United Nations Statistical Commission (UNSC) which established an Inter-Agency Expert Group for SDG indicators (IAEG-SDG). Under the auspices of the IAEG-SDG, various agencies were given custodianship for the finalisation of the appropriate indicators for the different SDG targets. FAO was given custodianship of 21 indicators across six SDGs (FAO, 2019).

SDG 2.4.1 is a set of tools used to measure concrete progress towards the achievement of SDG Target 2.4, one of eight targets under SDG 2, i.e. 'end hunger; achieve food security and improved nutrition and promote sustainable agriculture'. Specifically, Target 2.4 aims at 'by 2030, ensure sustainable food production systems and implement resilient agricultural practices that increase productivity and production, that help maintain ecosystems, that strengthen capacity for adaptation to climate change, extreme weather, drought, flooding and other disasters and that progressively improve land, soil quality' (FAO, 2019).

5.8.3.1 Characteristics of Indicator 2.4.1

By tracking the proportion of agricultural land area by its suitability status, SDG Indicator 2.4.1 provides an assessment of progress towards sustainable agriculture. In doing so, it supplies decision-makers with strategic information for evidence-based policies and action. The indicator is expressed by the following formula.

$$SDG\,2.4.1 = \frac{\text{Area under productive and sustainable agriculture}}{\text{Agricultural land area}}$$

This implies the need to measure both the extent of land under productive and sustainable agriculture (the numerator), as well as the extent of land area under agriculture (the denominator). The construction of the indicator must respect the following three conditions.

1. The indicator must reflect the priorities as they are expressed in the SDG target 2.4 and therefore consider issues related to resilience, productivity, ecosystem maintenance, adaptation to climate change and extreme events, and soils.
2. The preferred data source is the farm survey.
3. The need to define productive and sustainable agriculture implies the use of criteria to distinguish between sustainable and unsustainable areas.

Indicator 2.4.1 focuses on agricultural land, and therefore primarily on land that is used to grow crops and raise livestock. This choice of scope is fully consistent with the intended use of a country's agricultural area as the denominator of the aggregate indicator.

The following aspects are included within the scope of SDG Indicator 2.4.1.

- Both intensive and extensive production systems
- Subsistence agriculture
- State and common land when used exclusively and managed by the holding
- Food and non-food crops and livestock products
- Crops grown for fodder or for energy purposes
- Agroforestry (trees on the farm)
- Aquaculture, to the extent that it takes place within the agricultural area

The following aspects are excluded from the scope of SDG Indicator 2.4.1.

- State and common land not used exclusively by the agriculture holding
- Nomadic pastoralism
- Production from gardens and backyards, production from hobby farms
- Holdings focusing exclusively on aquaculture
- Forest and other wooded lands outside of an agricultural holding
- Food harvested from the wild

5.8.3.2 Sustainability themes and sub-indicators

Hayati (2017) identified a large number of potential sustainability themes across the three dimensions of sustainability and, for each theme, usually a large number of possible sub-indicators. The key considerations in the selection of themes are relevance and measurability. In terms of relevance, the relationship between the associated sub-indicator and sustainable agriculture outcomes at farm level should be strong. Following this approach, only sub-indicators that are responsive to farm level policies aimed at improving sustainable agriculture are considered. In terms of measurability, only a 'core' set of themes and sub-indicators for which measurement and reporting is expected in the majority of countries are selected. There are many relevant subthemes and sub-indicators. FAO proposes to measure Indicator 2.4.1 through a core set of 11 themes for global reporting purposes.

Selecting the most appropriate sub-indicator for each theme is a distinct step in the process. For any given theme, there may be multiple sub-indicators that are relevant and/or measurable. Consequently, in selecting the sub-indicators for Indicator 2.4.1 the following six key criteria have been considered (FAO, 2019).

1. **Policy relevance:** The indicator must be easily understood (reasons why it is selected) and the results easily interpreted by policy makers (is agricultural sustainability decreased and why? Which policies need to be implemented to address the issue?).
2. **Universality:** The indicator must be relevant for all countries in the world, both developing and developed.
3. **International comparability:** The way indicators are computed must ensure comparability across countries in order to ensure global reporting. Comparability, however, does not necessarily mean the use of absolute rate, even if these wage rates vary from one country to another. Similarly, compliance with national environmental standards or nationally recognised certification systems can be considered in computing environmental sub-indicators, even if national criteria vary from one country to another.
4. **Measurability:** Many themes are important sustainability issues but their measurement is difficult, complex or would involve costs that cannot be sustained in the framework of a regular monitoring exercise. To the extent possible, alternative measures have been proposed to maintain indicators that are considered relevant while offering feasible measurement solutions.
5. **Cost effectiveness:** Cost effectiveness is related to measurability. The cost associated with indicator measurement has systematically been considered in relation with the accuracy and reliability of the results obtained through different measurement options.

6. **Minimum cross-correlation between sub-indicators:** In selecting a limited set of themes and sub-indicators, efforts were made to reduce cross-correlation between different sub-indicators. High cross-correlation between sub-indicators would imply that two or more sub-indicators capture the same sustainability issue. In this case, the inclusion of one single sub-indicator, instead of several, would be sufficient to adequately measure agricultural sustainability performances.

The list of selected themes and sub-indicators is provided in table 5.5.

Table 5.5. List of themes and sub-indicators

Dimensions	S.No.	Theme	Sub-indicators
Economic	1.	Land productivity	Farm output value per hectare
	2.	Profitability	Net farm income
	3.	Resilience	Risk mitigation mechanisms
Environmental	4.	Soil health	Prevalence of soil degradation
	5.	Water use	Variation in water availability
	6.	Fertiliser pollution risk	Management of fertilisers
	7.	Pesticide risk	Management of pesticides
	8.	Biodiversity	Use of biodiversity-friendly practices
Social	9.	Decent employment	Wage rate in agriculture
	10.	Food security	Food insecurity experience scale
	11.	Land tenure	Secure tenure rights to land

Source: FAO, 2019

For each sub-indicator, criteria to assess sustainability levels are developed. The concept of sustainability implies an idea of continuous progress and improvement towards better performances across all themes and such performances can therefore be more or less sustainable. In order to capture the concept of continuous progress towards sustainability, three sustainability levels such as desirable, acceptable and unsustainable are considered for each sub-indicator.

SDG Indicator 2.4.1 measures progress towards more sustainable and productive agriculture. For many sub-indicators, it is likely that changes will be relatively limited from a year to another. It is therefore recommended that the survey be conducted every three years. The survey's sampling design must respond to the need to capture the structure and the different typologies of agricultural holdings. The possible instruments/ sources of information for each sub-indicator are furnished in table 5.6.

Table 5.6. Possible data collection instruments for each sub-indicator

S.No.	Sub-indicators	Possible data collection instruments
1.	Farm output value per hectare	Agricultural surveys, household surveys linked with administrative records and market surveys, remote sensing, agricultural and livestock census
2.	Net farm income	Agricultural surveys, household surveys linked with administrative records and market surveys, agricultural and livestock census
3.	Risk mitigation mechanisms	Household surveys with agricultural information, community surveys, administrative records
4.	Prevalence of soil degradation	Environmental monitoring systems, soil sampling, remote sensing calibrated with ground observations, GIS data/ maps/ models calibrated with ground observations and samplings
5.	Variation in water availability	River flows records, water level records, abstraction records, remote sensing, GIS information/ maps/ hydrogeological models, administrative sources, household surveys
6.	Management of fertilisers	Environmental monitoring systems (soil, water quality), agricultural surveys, GIS data/maps and models based on sale data, agricultural surveys and administrative sources
7.	Management of pesticides	Environmental monitoring systems (soil, water quality), agricultural surveys, models based on active substance sale data, agricultural surveys and administrative sources
8.	Use of biodiversity-friendly practices	Environmental monitoring systems including remote sensing (land use/ land cover), GIS data/ maps
9.	Wage rate in agriculture	Labour force survey, household survey with agricultural module, administrative data
10.	Food insecurity experience scale	Household surveys, health data
11.	Secure tenure rights to land	Household surveys with agricultural module, administrative/ legal sources

Source: FAO, 2019

6

Low External Input and Sustainable Agriculture

Low external input agriculture developed as a response to the criticisms surrounding Green Revolution technology being neither sustainable nor feasible for many small-scale farmers around the world. The concept aimed to redesign the agricultural system by optimising the use of biological resources, keeping changes to the natural ecosystem as well as the use of external inputs at a minimum (Pimentel et al., 1989).

6.1 CONCEPT OF LEISA AND HEIA

Low external input agriculture (LEIA) is described as an approach referring to a set of agronomic practices that aim to reduce the use of inputs from outside the production system. These inputs may include water, energy, seeds, chemicals, etc. According to the FAO, LEIA does not mean the total elimination of external inputs, such as pesticides and fertilisers; rather it gives more emphasis on improved agronomic practices, integrated pest management, labour and overall farm management, towards maintaining yields (Corsi and Muminjanov, 2019). LEIA is an approach that can be applied to various production systems in various geographical contexts. The other terms in use with the same concept as LEIA include low external input sustainable agriculture (LEISA) and low input farming systems (LIFS).

One of the most promising paradigms that has emerged for the benefit of small scale resource-poor farmers is low external input and sustainable agriculture, which can enable such farmers to achieve higher income and attain sustainability by following two considerations (Gold, 2007).

1. Optimising the use of locally available resources, thereby achieving a synergetic effect among the various components of the farming system (soil, water, animals, plants, etc.) so that they complement each other in the production of output.
2. Minimising the use of external inputs, except where there is a serious deficiency and where the effect on the system will be to increase recycling of nutrients.

Thus, the low external input sustainable agriculture (LEISA) is a way of farming, where it seeks to optimise the use of locally available resources by maximising the complementary and synergistic effects of different components of the farming systems. External inputs are used in a complementary way. Here the aim is not to maximise short-term production, but to attain an adequate and sustainable level over the long-run. To achieve these goals, LEISA must tap the most viable indigenous knowledge and practices and ecologically friendly technologies in a given ecological and sociocultural setting, since the experience in one agroecological setting may not be appropriate in other areas (FAO, 1996).

The practice of LEISA is guided by the following four basic principles.

1. Securing favourable soil conditions for plant growth particularly managing organic matter and enhancing soil life.
2. Optimising the nutrient availability and balancing the nutrient flow, particularly by means of nitrogen fixation, nutrient acquisition and complementary use of external fertilisers.
3. Minimising the losses due to plant and animal pests by means of prevention and safety treatment.
4. Minimising losses due to flows of solar radiation, air and water by way of microclimate management, water management and erosion control.

In the high external input agriculture (HEIA), external inputs like chemical fertilisers, high yielding and hybrid seeds, pesticides, irrigation, mechanisation based on fossil fuels are extensively used. However, here the problem is related to excessive and unbalanced use of agrochemicals, salinisation from irrigation, and loss of plant diversity which can in the long run harm humans, animals, and the soil. Increasingly, HEIA areas are experiencing a decrease in the effectiveness of these external inputs because of biological factors and also because of a loss in profitability as the soil productivity is reduced over time. But HEIA could rapidly increase agricultural production to meet the demand for food for the increasing population. The other benefits of HEIA include

- Many problems related to diseases caused by mal-nutrition and deficiencies can be effectively managed.
- New improved varieties give yields within a short period of time.
- Mechanisation solves the problem of labour shortage.
- Income and profit margins of the products may be increased.
- Productivity of land is improved.
- A market facility for production is increased.

With all the above benefits, the high external input agriculture on long-run developed serious threats mostly on ecological and environmental issues.

- The environmental balance is collapsed due to lack of biodiversity by planting a few cash crops.
- Soil erosion is increased due to constant furrowing by machinery.
- Dependence on imported machinery, chemical fertiliser, pesticides, hybrid seeds and other inputs is increased.
- Extensive use of pesticides disturbs the natural mechanism of controlling pest and diseases as the artificial pesticides kill both pests and their natural enemies.
- Use of agrochemicals adversely affects the soil *p*H, cation exchange capacity, soil structure, soil texture and soil organisms that consequently lead to reduced microbial activities in the soil.
- The large farmers get benefited with high capital investment while small-scale farmers who were short of capital run into debt.
- Traditional varieties of seeds and their genetic resources face extinction due to introduction of hybrids.
- Conventional agricultural knowledge and techniques are neglected and extinguished.

Table 6.1. The characteristics of HEIA and LEISA

Characteristics of HEIA	Characteristics of LEISA
1. The farming pattern depends heavily on external and chemical inputs. Although yields have increased substantially, contributing to raising total production, farmers and the environment have had to pay the price for keeping up with this development.	LEISA relies on the optimal use of natural processes.
2 The focus of agricultural development and research has mainly been on maximising yields coupled with increasing specialisation of production.	The focus is on the sustainability of farming system.
3 There is a great damage to the environment.	This is environmentally sound and has the potential to contribute to the long-term sustainability of agriculture.

4	The continuing drop in prices of farm produce and the rising costs of agricultural inputs make farming increasingly unprofitable.	LEISA puts greater emphasis is on the long-term sustenance and promotes a balance between the profit and livelihood.
5	HEIA depends on the higher production and profit, without consideration of the local needs and local market.	Sustainable ecological practices depend largely on local agroecolo-ical conditions and on local socioeconomic circumstances, as well as on farmers' individual needs and aspirations.
6	Primarily one or two commodity driven development, lack of diversity in the farming practices; as a result, there is greater risk of failure and price fluctuation. The number of products and commodities are very less.	One way of LEISA is to diversification of farms; with a range of crops and/or animals, farmers will suffer less from price fluctuations or drops in yield of single crops. Maintaining diversity will also provide a farm family with a range of products to eat or sell throughout a large part of the year.
7	Under HEIA system, soil quality deteriorates, and there is resurgence of pests, lack of resilience in the soil-plant system.	LEISA maintains a healthy soil, recycling nutrients on the farm, and utilising approaches such as integrated pest management.
8	In HEIA, there is lack of use of indigenous technologies.	Best technologies, for example, soil and water conservation (terraces, ditches, and vegetation strips on sloping land), better timing of operations, improved crop spacing and densities, manure or compost and water application based on local conditions are used to increase the productivity.

6.2 CRITERIA FOR LEISA

Important characteristics of LEISA systems are that they are based on a preventive approach whereby the problem is tackled at its roots, as opposed to the more symptom-curing nature of modern high input agriculture. Ecological and biological principles are the basis of the farm system. Nature works for the farmer, the farmer does not have to work against it. They are often based on local knowledge and production systems adapted to modern requirements and much less on external expertise. They are generally more labour-intensive compared to the mechanisation- and chemical-intensive character of the modern agriculture.

Ecological criteria

1. Balanced use of nutrients and organic matter
2. Efficient use of water resources
3. Diversity of genetic resources
4. Efficient of genetic resources
5. Efficient use of energy sources
6. Minimal negative environmental effects
7. Minimal use of external inputs

Economic criteria

1. Sustained farmer *livelihood* systems
2. Competitiveness
3. Efficient use of production factors
4. Low relative value of external inputs

Social criteria

1. Wide-spread and equitable adoption potential, especially among small farmers
2. Reduced dependency on external institutions
3. Enhanced food security at the family and national level
4. Respecting and building on indigenous knowledge, beliefs and value systems
5. Contribution to employment generation

6.3 ECOLOGICAL PRINCIPLES OF LEISA

The principal objective of LEISA is to maintain the agricultural production at an optimum level using less external inputs in an eco-friendly environment. To achieve this objective, the LEISA practices are concentrated heavily on maintaining a living soil, creating biodiversity, recycling of resources and natural pest management. Basic ecological principles of low external input sustainable agriculture are discussed below.

6.3.1 A Living Soil

Soil can be regarded as a non-renewable resource, as soil formation is a very slow process. The soil provides a medium to anchor plant roots, but is also a very complex ecosystem. The climate, animals, plants and human beings influence the physical, chemical and biological characteristics of the soil. Adequate amount of water, air and nutrients in the soil is essential to maintain crop production at a sustainable level. Favourable soil structure is essential to retain water, nutrients and the growth of root systems of the plants. It is important that soil should be free from poisonous substances.

A productive agricultural soil is full of life, with millions of microorganisms which all interact chemically and physically with their soil environment. These processes regulate the release of nutrients from minerals and organic matter to feed the plants. A living soil has a better structure and can absorb and retain more water and air than a sterile soil. Sound ecological production therefore begins with improving the soil.

The following practices can help improving the condition of the soil.

1. Growing legumes to fix atmospheric nitrogen and provide it to the crop
2. Feeding the soil with as much organic matter as possible through green manure, compost, cover crops, crop residues and agroforestry practices
3. Keeping the soil covered at all times with mulch or cover crops
4. No or reduced tillage which enhances water availability and soil conservation
5. Cultivating a range of crops and animals to reduce risks of disease and pest outbreaks, maintain a balanced nutrient supply and provide resilience
6. Planting trees on contours or making terraces to prevent soil erosion by wind or water

6.3.2 Biological Diversity

The diversity of different species of plants and animals, and the genetic variation within each species, provides the vital resource of biological diversity, which enables life on earth. Healthy ecosystems are relatively stable and the diversity they contain enables them to adapt to changing circumstances. For many small farmers the available agrobiodiversity is the basis of survival. A mix of different locally adapted crops and animals and different varieties of the same increases on-farm diversity, increasing the chances of producing something even under adverse conditions. These principles of traditional farming can be further developed and used systematically in ecological farming (LEISA, 2006).

1. Intercropping in time and space: Planting different crops together in different combinations or formations, or in sequence, can optimise the use of available resources and reduce the pressure of pests.
2. Different plant species can also be used to support the ecological functioning of the whole farming system: Trees or bushes for windbreaks, flowering plants which provide food and habitat for beneficial insects that help control

pests, shade trees for light-sensitive plants, trees to provide green manure and fuel wood.

3. Integration of different crops or weeds with animals for better utilisation of resources: Fish in rice fields, integrated crop-chicken-fish systems and other combinations of crops and animals.

6.3.3 Water

Growing population, rapid urbanisation and increasing industrial and agricultural production are all increasing competition for and pressure on water. As agriculture is one of the major users of water, and one of the major polluters of water resources, it is imperative that water use in agriculture is as efficient as possible and that leaching of surplus nutrients and in small scale farming it is important to make the best possible use of the limited amounts of available water. Infiltration can be improved by keeping the soil covered, through minimum disturbance of the soil, adding organic matter from cover crops and mulching.

When introduced, water-harvesting systems are generally multipurpose. Farm ponds, earth dams and subsurface tanks will often serve as a source of drinking water and water for livestock during periods of water scarcity. Water harvesting can open up new livelihood options. Subsistence farmers, who invest in water harvesting systems with a storage component, often diversify their farming system to include cash crop production, for the local market during off-season when prices are high. This diversification increases the resilience of farm households, as they are better equipped to cope with periods of climatic hazards such as droughts and floods.

6.3.4 Energy

Solar energy is captured by plants that are able to transform it into biomass. This is the basis for all higher life forms, animals as well as humans, and is a process that is unique for green plants. Biomass contains stored energy as well as nutrients, and agriculture should focus on maximising the amount of solar energy, which is captured and transformed into plant growth and thereby food and fodder resources. But additional energy is required for cooking and heating and is useful for irrigation, threshing and processing. There are many opportunities to make use of renewable energy, most frequently fuel wood, straw, crop residues and even manure are used. There are also other possibilities to make use of renewable energy; small scale bio-digesters which use manure, solar energy devices, small scale hydropower generators, wind-power,woodlots for fuel wood.

6.3.5 Animal-Plant Interaction

In nature, nothing functions in isolation; everything depends on the other factors present. In animal production, to optimise the performance of cattle, it is very important that management practices should enhance the ecological functioning of the web of living organisms within the production system - climate, soil and soil life, vegetation and cattle - by influencing their interactions.

For cattle production, it is important that the breed is selected first, then the pasture suited to that breed and finally the soil is corrected with proper fertiliser or amendment to make the pasture grow. This order has to be reversed. The pasture has to be adapted to the soil and the cattle to the pasture, and all of it has to fit the climate. In addition, the forage crops are to be grown. Of course, in dry areas, forage yields depend strongly on the availability of water. In a well-structured soil, roots are able to explore a larger soil volume for more water and nutrients. Integrating deep-rooting crops and trees into the pasture system will further increase the production of biomass, the overall performance of the system.

In native grassland, cattle always first eat the plants it likes most. The plants that are not eaten get old, hard and are not tasty. The eaten plants sprout again and are grazed on another time. This goes on until these palatable plants disappear. But the less appreciated plants continue to grow and multiply and with time the entire pasture gets hard, rough and has little nutritive value. Then the ranchers set fire to the pasture. Many plants die, and only those that can protect their growing points against fire survive. Then the pasture becomes worse and the forage volume smaller. Thus, all organic matter that nourishes soil microorganisms is burned out, resulting in their death. The soil compacts, water runs off and the vegetation gets scantier. Thus, the division of pastures into smaller subunits for grazing rotation is fundamental for sustainability of pasturelands.

However, to prevent 'global climate change' by 'greenhouse gases', it is important to reduce methane emission by cattle. This obliges farmers to speed up animal production per unit area and to reduce the slaughter age to get a lower ratio of kg methane/kg animal protein. The use of grains for animal feed has to be reduced as well, giving priority to human consumption. This then increases the dependence on forage. But, as grass cellulose is the main source of methane emission, management practices that contribute to an increase of forage yield per unit area and maintain stocking rate without weight losses are needed.

6.3.6 Local Resources-based Integrated Crop-Livestock Systems

The present livestock production systems in most industrialised countries are in direct competition with human needs. Livestock presently consume almost 50%

of world cereal grain supplies. In the intensive large-scale production systems, increasingly promoted by corporate agriculture, livestock wastes contaminate soil and water resources, create less than favourable working conditions for the personnel involved in feeding and cleaning, and decrease employment opportunities. To meet food needs in 2050, it is necessary to develop livestock production systems, which do not depend on cereal grain.

In developing countries like India, instead of grain-based livestock systems, alternative production systems must be developed which make optimal use of locally available resources. Close integration of livestock in the farming system, with recycling of all excreta, will be the basis of agriculture, which can be highly productive and also sustainable. In tropical countries, especially in the humid zone, there are many crops and farming systems that considerably exceed the productive capacity of grain cereals. Key plants in this scenario are sugarcane, cassava, and the palm family, especially the oil and sugar palms.

6.4 PROMISING LEISA TECHNIQUES AND PRACTICES

Farming techniques encompass all human activities on a farm that aim at enhancing agricultural production. These activities can involve skilful management of farm resources, assets, inputs and/or outputs. They combine human knowledge, insights and skills with technical means and are mainly oriented towards managing the physical and biological components and processes of the farm (Reijntjes et al., 1992). A technology system involves the whole complex of techniques used in a farm system. A LEISA technology system is a combination of deliberately chosen techniques oriented towards sustainability. LEISA techniques have productive, protective, reproductive and/or social functions, and complement each other. Under the complex and diverse conditions of LEISA farmers, techniques are farm-specific and depend on the local availability of skills, assets and inputs. In view of the often high variability of climatic conditions, an important characteristic of LEISA systems is flexibility in the choice of techniques. Promising LEISA techniques and practices are those which farmers in specific areas have found to be effective in making their farm systems more sustainable (Reijntjes et al., 1992).

6.4.1 Use of Mineral Fertiliser

Nutrient management is managing the amount, source, placement, form, and timing of the application of nutrients and soil amendments to ensure adequate soil fertility for plant production and to minimise the potential for environmental degradation, particularly water quality impairment. Soil fertility traditionally dealt with supplying and managing nutrients to meet crop production requirements, focusing on optimisation of agronomic production and economic returns to crop production. Contemporary nutrient management deals with these same production concerns, but recognises that ways of farming must now balance the limits of soil and crop nutrient use with the demands of intensive animal production.

Many tropical soils are very poor in plant nutrients or have specific nutrient deficits and crop growth is generally poor in these soils. With the harvest of crop plants nutrients are also removed permanently from the soil. Thus, to maintain or enhance the yield level, external inputs of nutrients are required to replenish the removed nutrients by harvesting of previous crops. Like plants, soil microbes and soil animals also need mineral nutrients. Mineral fertiliser normally increases the availability of biomass which in turn, increases the deposition of organic manure in the soil. This may enhance soil life. On the other hand, enhanced soil life increases the efficiency of mineral fertilisers. Application of mineral fertilisers in low to moderate dose with organic manures and micronutrients enhance soil balance, nutrient availability and, hence, the level and sustainability of crop production. The efficiency and recycling of fertiliser can be further increased by controlling weeds, pests, diseases, erosion and leaching; by rotating shallow- and deep-rooting crops; by split doses of nitrogen and by applying it below the soil surface and near the root zone.

Continuous use of easily soluble concentrated mineral fertiliser can have harmful effects on soil. This may disturb soil life and lead to acidification, micronutrient depletion, soil degradation, poor crop health and lower crop yields (Reijntjes et al., 1992). Ammonium sulphate is a very strong biocide which hinders nitrogen fixation and kills nematodes and earthworms. Among mineral fertilisers, calcium ammonium nitrate is less harmful than urea, which is still to be preferred over ammonium sulphate. This is because urea and ammonium sulphate acidify quickly, the later being quicker, leading to high levels of toxic soluble aluminium in the soil (Smaling, 1990). Mineralisation of rock phosphate can be accelerated in acid conditions. Rock phosphate can be composted together with manure and plant residues for better result. Rock phosphate may also be applied to legumes to enhance nitrogen fixation. The choice of mineral fertilisers will depend on soil type and other variables. It is always better to go for soil testing which will help farmers find the right kind and amounts of mineral and organic fertilisers. The availability and price of mineral fertilisers may limit their use. Appropriate and balanced levels of application can improve the economics of using mineral fertilisers.

6.4.2 Use of Organic Manures

Organic manures are organic materials derived from animal, human and plant residues which contain nutrients in complex organic forms. They are the sources of plant nutrients. They release nutrients after their decomposition and provide organic acids that help to dissolve soil nutrients and make them available for the plants. Organic manures can be grouped into bulky organic manures and concentrated organic manures based on the concentration of the nutrients. The organic manures provide several benefits.

1. Organic manures supply plant nutrients including micronutrients needed for optimum plant growth.
2. Continued use of manures builds organic matter in soils and improves soil structure. This modification of soil structure helps improve water holding capacity, aeration, friability and drainage.
3. Organic manures improve soil condition for better penetration of roots into deeper layers.
4. Organic manures increase the availability of nutrients through improvement in cation exchange capacity.
5. Carbon dioxide released during decomposition acts as a CO_2 fertiliser.
6. Organic manures increase soil microbial biomass carbon which improves rhizosphere environment.
7. Plant parasitic nematodes and fungi are controlled to some extent by altering the balance of microorganisms in the soil.

But in many developing countries, locally available valuable organic resources are not used to their fullest advantage for agriculture. This is because of inefficient handling and use of organic farm by-products for the purposes other than manures. Improved collection, composting, storage and transport of dung and urine can reduce nutrient losses. The nutrient content of manure is directly related to the species, sex and age of the animals and the quality and quantity of the feed and bedding material. The quality and quantity of manure can be improved by choosing appropriate animal species and adjusting their numbers to the available feed resources; and by balancing the protein/energy content in feed and providing good bedding material and housing for livestock. Nutrient cycling within a farm system can also be improved through the use of manure from poultry. By making simple night-housing for backyard poultry that run freely by day, smallholders can collect manure, which can then be applied on small plots.

Farmers in the hill agriculture follow a system of close integration of crop, livestock and forestry/grassland management. Farmers maintain traditional practices such as terracing, manure management, legumes intercropping, and mulching where appropriate. Farm yard manure quality can be improved by better decomposition and the N-content can be increased through proper management of urine and manure. Liquid manure can be prepared from urine and various plant extracts rich in minerals or secondary plant compounds. These 'manure teas' have been found to be effective liquid fertilisers on crops such as vegetables and also for organic pest and disease management. The use of urea fertiliser is expected to decline in the areas where this liquid manure is used.

Organic manures are not free from some limitations. Nutrients from organic manures are not immediately available to the plants. They are released slowly and over a longer period of time than from most commercial fertilisers. So if there is an immediate need for nutrients, organic manures cannot readily supply nutrient to plants. Many organic manures have low nutrient content and therefore need to be applied in larger quantities to obtain sufficient quantities of nutrients for crops. If there are large distances between the places where dung, feed and bedding material are collected, the places where manure is stored and the fields where it is applied, manure handling will require much time and energy. Handling and transportation cost becomes more. This may also be labour-intensive. Some organic manures need composting before its application to the field. Improperly processed organic manures may contain harmful pathogens and may also contaminate produce. The use of fresh manure may introduce new weeds into fields since certain weed seeds remain alive even after passage through animals and composting process. Cultural taboos can make handling of manure unacceptable to some people.

6.4.3 Crop Residue Management and Conservation Tillage

Conservation tillage is a term that covers a broad range of soil management systems that leave residue cover on the soil surface, substantially reducing the effects of soil erosion from wind and water. These practices minimise nutrient loss, decreased water storage capacity, crop damage, and improve soil quality. The soil is left undisturbed from harvest to planting except for nutrient amendment. Weed control is accomplished primarily with herbicides, limited cultivation, and in more sustainable systems, with cover crops. At least 30% of crop residues must be left in the field to qualify for conservation tillage system. Some specific types of conservation tillage are minimum tillage, zone tillage, no-till, ridge-till, mulch-till, reduced-till, strip-till and rotational tillage.

6.4.4 Agricultural Waste Management

There is nothing such as organic waste in a farming system. The so called farm wastes such as green leaves, dry leaves, weeds, cow dung and urine, fodder wastes from cattle sheds, cow dung gas slurry, etc. can be converted to valuable organic manures. These items are put in alternate layers (cow dung in between layers) in a heap and left for 45 days. The heap is turned once or twice. In 45 days, the items get semi-decomposed. This partially decomposed material can be further used in two ways. By releasing earthworms to this material, vermicompost can be prepared in another 45 days. The other way is to add coir pith, tank silt and quarry dust to semi-decomposed material, which turns into good quality manure in 45 days.

6.4.5 Green Manuring and Cover Crops

Trees, shrubs, cover crops, grain legumes, grasses, weeds, ferns and algae provide green manure, an inexpensive source of organic fertiliser to build up or maintain soil organic matter and fertility. The cumulative effects of continued use of green manures are important, not only in terms of nitrogen supply but also with regard to soil organic matter and phosphate and micronutrients which are mobilised, concentrated in the topsoil and made available for plant growth. Green manuring adds organic matter to the soil and increases the availability of certain plant nutrients like P, K, Ca, K, Mg and Fe. Leguminous crops add nitrogen to soil through atmospheric nitrogen fixation. Green manuring improves the soil structure and other soil physical properties. Deep-rooted green manure crops in a rotation can help recover nutrients leached to the subsoil. Green manure crops hold plant nutrients that would otherwise be lost by leaching. The practice of green manuring also facilitates infiltration of water thus decreasing runoff and soil erosion. Rapid establishment of a cover crop is crucial to its potential to suppress weeds. If a cover crop is being introduced into a crop rotation, good potential to suppress weeds is a key to farmer acceptance, as beneficial effects from weed control can be observed immediately, while effects of improved nutrient supply may occur only in the longer run.

Green manure crops can be planted in different combinations and configurations in time and space.

1. **Improved fallow:** Natural fallow vegetation is replaced with green manure crops to speed up regeneration of soil fertility and permit permanent cultivation. These green manures may be left to grow for one or several years, or only during the dry season.

2. **Alley cropping:** It is a form of simultaneous fallow in which fast-growing trees or shrubs, usually legumes are planted in rows and are regularly cut back. The prunings are used as mulch or worked into the soil in the alleys between the rows.

3. **Integration of trees into cropland:** Tree legumes are grown among the crops and are regularly cut for mulch material to maintain soil fertility and conserve soil moisture. For example, *Erythrina poeppigiana* is grown in coffee plots.

4. **Live mulching:** In this practice, the rows of food crops are sown into a low but dense cover crop of grasses or legumes, strips of the cover crop are removed by hand or killed by herbicides when the food crops are to be sown, thus reducing soil tillage operations to zero.

Green manure and cover crop species should fit the agroecological condition. In general, these crops should have the characteristics such as easy establishment, vigorous growth under local conditions, ability to cover weeds quickly, ability to either fix atmospheric nitrogen or concentrate plenty of phosphorus, should also have multiple uses.

Various nitrogen-fixing leguminous and non-leguminous species, particularly trees, creepers and bushes can be used as green manures. Using grain legumes for green manuring brings quick economic benefit but, as they tend to accumulate nutrients in the grain, which is then harvested, their positive effect on subsequent crop yields is usually low. The green manure and cover crops can be grown intercropped with another food crop. Mixtures of green manure crops are often more successful than sole crops, as they are less susceptible to pest attacks and combine different characteristics needed for improved fallow, such as quick soil cover and deep rooting. As legume growth depends on the presence of suitable *Rhizobium* strains, inoculation may be necessary. Applying organic manures improve quick establishment of legume. Tropical grasses such as *Pennisetum purpureum* or *Panicum maximum* can produce much biomass and accumulate phosphorus and potassium more quickly than many legumes.

The main problem in adoption of green manuring is that the land on which green manure crop is raised could have been profitably utilised for growing a crop of economic importance. Another problem is adequate soil moisture either through rainfall or irrigation is essential for *in situ* decomposition of green manure crop. Under rainfed condition, if sufficient rainfall is not received proper decomposition may not take place, and the germination of the succeeding crop is hampered. There is a possibility of incidence of diseases and pests, even nematodes.

6.4.6 Mulching

Mulch can be defined as a shallow layer at the soil/air interface with properties that differ from the original soil surface layer. Any material, such as straw, plant residues, leaves, loose soil or plastic film which is placed on the soil surface to reduce evaporation, runoff, erosion or to protect plant roots from extremely low or high temperature, to suppress weed growth is called as mulch; and the practice of applying mulches is called mulching.

Mulching has the following advantages.

1. Mulching favourably influences the soil moisture regime by controlling evaporation from soil surface, improving infiltration and soil moisture retention and facilitating condensation of water at night due to temperature reversals.
2. Mulching suppresses weed growth.
3. Mulching invariably decreases soil erosion and often reduces runoff rate. Mulch cover protects the soil from raindrop impact and surface sealing, increases the infiltration rate and decreases runoff velocity through physical resistance to water flow.
4. Mulch has a moderating influence on the soil thermal regime.
5. Crop residue mulch improves soil aeration by promoting free exchange of gases between the soil and the atmosphere.
6. Mulching improves soil structural properties directly by preventing the raindrop impact and indirectly by promoting the biological activity.
7. Organic mulches add organic matter and plant nutrients to soil upon decomposition. Thus, they improve carbon sequestration. Cation exchange capacity is substantially influenced by organic matter content in soils containing predominantly low activity clays.
8. Soil biological activity is either directly influenced by supply of food substrates by organic mulches or indirectly influenced by both organic and inorganic mulches through alteration of soil hydrothermal regime.
9. Mulching reduces the need for tillage.

The mulch materials include organic residues (grass clippings, leaves, hay, straw, shredded bark, sawdust, wood chips, shredded newspaper, cardboard, wool, etc.), compost, rubber mulch, plastic mulch, organic sheet mulch (various products such as biodegradable alternative to plastic mulch), rock and gravel mulch, living mulch, etc. The effects of mulch depend on its composition and colour, the amount applied, the timing of application and the rate at which the mulch decomposes. This rate depends, in turn, on the form and timing of application and the meteorological conditions in the soil and air. Mulch is usually applied towards the beginning of the growing season, and may be reapplied as necessary. It serves initially to warm the soil by helping it retain heat. This allows early seeding and transplanting of certain crops in cold regions, and encourages faster growth. As the season progresses, the mulch stabilises temperature and moisture, and prevents sunlight from germinating weed seed.

Possible constraints to mulching are insufficient availability of mulch material, unsuitable soils, soil compaction and pest problems (rodents, insects, fungi, persistent weeds). The infestation of termite is aggravated due to mulching especially in the humid areas. There are also some problems of mulching with respect to its adoption by the farmers. Mulch hinders the operation of sowing (if it is applied before sowing), fertiliser applications, irrigation, etc. Sometimes thick mulching in areas of high rainfall may lead to waterlogging conditions and provide an environment for disease infestation. Plastic mulches usually require pick up and disposal at the end of the season and their manufacture and disposal entail significant environmental cost, as these are not biodegradable. Plastic mulch can interfere with the recharge of soil profile by rain or overhead irrigation.

6.4.7 Preventing Land Degradation

A good plant cover is very important for preventing soil degradation and achieving soil rehabilitation. A vegetative cover has a number of beneficial effects on soil aeration, soil moisture and organic matter content, physical characteristics and biological activity in the soil. In addition, a plant cover protects against soil erosion. Cover crops can be quite aggressive creepers and may compete with the main crop. On the other hand if leguminous plants are used as cover crops they add nutrients because they fix atmospheric nitrogen and make it available for the crop. Selection of crops is very important while dealing with degraded soils.

6.4.8 Integrated Pest Management

Integrated pest management (IPM) is an ecologically based approach to pest control that utilises a multidisciplinary knowledge of crop/pest relationships, establishment of acceptable economic thresholds for pest populations and constant field monitoring for potential problems. The following LEISA management practices may be included in IPM (LEISA, 2006).

1. Use of resistant varieties
2. Crop rotation
3. Cultural practices
4. Optimal use of biological control organisms
5. Use of certified seed
6. Protective seed treatments
7. Disease-free transplants or rootstock
8. Timeliness of crop cultivation
9. Improved timing of pesticide applications
10. Removal of infested plant material

The term biointensive IPM emphasises a range of preventive tactics and biological controls to keep pest population within acceptable limits. Reduced risk pesticides are used if other tactics have not been adequately effective, as a last resort and with care to minimise risks. Biological control is, generally, man's use of a specially chosen living organism to control a particular pest. This chosen organism might be a predator, parasite, or disease, which will attack the harmful insect. A complete biological control programme may range from choosing a pesticide which will be least harmful to beneficial insects, to raising and releasing one insect to have it attack another, almost like a 'living insecticide'.

6.4.9 Windbreaks

Windbreaks are barriers used to reduce and redirect wind. They usually consist of trees and shrubs, but may also be perennial or annual crops and grasses, fences, or other materials. The reduction in wind speed behind a windbreak modifies the environmental conditions or microclimate in the sheltered zone. The direction from which wind is blowing is called windward side and direction to which wind is blowing is called leeward side. Besides influencing the microclimate, hedges can be useful in keeping animals out of fields and/or producing fruits, herbs, fodder, mulch, thatching material or fuel. They also play a role in balancing pest populations. The improvement in microclimate as a result of windbreaks can lead to an increase in crop yield and may compensate for the land lost to planting trees. The total biomass production of both crops and hedges may be raised considerably.

Height, density, number of rows, species composition, length, orientation, and continuity determine the effectiveness of a windbreak in reducing wind speed and altering the microclimate. Windbreaks are most effective when its orientation is such that it is at right angles to prevailing winds. The purpose and design of each windbreak is unique, thus the orientation of individual windbreaks depends on the design objectives. To control soil erosion, windbreaks should be planted to block the prevailing winds during the times of greatest soil exposure. Constraints to planting windbreaks may be excessive competition for light and root space with the adjacent crop, and pest transmission from plants in the hedge to certain crops. Establishment may be difficult because of long dry periods and free-roaming animals. Allowing natural regrowth of trees on the borders of fields or contour lines can be a viable alternative to deliberately planting trees.

6.4.10 Water Harvesting

In rainfed agriculture, efficient water management is of great importance. Rainfall may be too low or too high or too irregular, creating high risks of yield losses and/or unfavourable growth conditions or damage by erosion. Where there is not enough rain to grow a crop or where rainfall is very irregular, water-harvesting techniques can be used to concentrate rain or flood water in such a way that it can be used for crop growth. Thus, water harvesting, defined in its broadest sense as the collection of runoff for its productive use. More specifically, it is the collection and storing of water on the surface of the soil for subsequent use. It is relevant to areas where the rainfall is reasonably distributed in time, but inadequate to balance potential evapotranspiration of crops. Water harvesting supports a flourishing agriculture in many dry areas. The sustainability of the various water harvesting techniques is found to depend largely upon the timing and the amount of rainfall. However, the techniques should be site-specific. Water harvesting is mainly practised in arid and semiarid regions, where surface runoff often has an intermittent character. Water harvesting is based on the utilisation of runoff and requires a runoff producing area and a runoff receiving area. Because of the intermittent nature of runoff events, storage is an integral part of the water harvesting system.

Water harvesting not only secures and increases crop production in regions where rainfall is normally insufficient; it can also serve to control soil erosion and to recharge aquifers tapped for irrigation. An additional benefit is the improvement in soil fertility. Silt, manure and other organic matter are harvested together with the water. The soil profile stays moist longer, stimulating soil life and improving humus formation, nutrient availability and the soil's capacity to hold water.

Water may be stored directly in the soil profile or in small reservoirs, tanks, and aquifers. Water harvesting system should therefore have the following four components.

1. Runoff producing catchments
2. Runoff collection scheme
3. Runoff storage facility
4. Cultivated or cropped area

The water harvesting structures which are commonly built for surface storage and/or groundwater recharge are check dams, percolation tanks, subsurface dykes, contour bunding, gully plugging, etc.

Water ponds/tanks: This is the most commonly used method to collect and store rainwater in ponds or tanks. Most ponds have their own catchments, which provide the requisite amount of water during the rainy season. Where the catchments are too small to provide enough water, water from nearby streams is diverted through open channels to fill the ponds. In some places water from irrigation canals is also used to fill ponds. Ponds are excavated in different shapes and sizes depending upon the nature of the terrain, availability of land, water requirements of the village community, etc.

Check dams: These are concrete or masonry structures built across small streams for surface storage and incidental benefit of groundwater recharge. The design of these structures is done considering the volume of water that can be stored in the upstream channel.

Percolation tanks: Percolation tank is an artificial reservoir which is constructed across stream, submerging a land area with adequate permeability to facilitate sufficient percolation to collect surface water runoff and allow it to percolate within the permeable land. This is one of the effective methods of refilling groundwater table. Moderate to high porosity of soil and/ or underlying rocky strata is the main criteria for the choice of percolation tanks. Percolation tanks should normally be constructed in a terrain with highly fractured and weathered rock for speedy recharge. Too high permeability may result in the percolated water escaping in the downstream as regenerated surface flow which will defeat the purpose of water conservation. A long-term evaluation of the pattern of rainfall for the benefitted area must be studied so that the percolation tank gets filled up fully during monsoon. The amount of water a percolation tank would be able to seep into the ground for compensating depleted groundwater, depends upon the amount and nature of precipitation of the area, thickness of topsoil and weathered zone, type of vegetation, evaporation from the surface of wet soil, profile of underlying hard rock, the topographical features of the sub-basin, and the status of soil.

Subsurface dykes: Subsurface dyke or underground dam is a subsurface barrier across stream which retards the base flow and stores water upstream below ground surface. By doing so, the water levels in upstream part of groundwater dam rises saturating otherwise dry part of aquifer. The site where subsurface dyke is proposed should have shallow impervious layer with wide valley and narrow outlet. After selection of suitable site, a trench of 1-2 m wide is dug across the breadth of stream down to impermeable bed. The trench may be filled with clay or brick/ concrete wall up to 0.5 m below the ground level. Since the water is stored within the aquifer, submergence of land can be avoided and land above the reservoir can be utilised even after the construction of the dam. No evaporation loss from the reservoir and no siltation in the reservoir take place. The potential disaster like collapse of the dams can also be avoided.

Contour bunding: These are small earthen bunds built horizontally in parallel rows across the hill slope to collect runoff from the catchment to improve soil moisture on the cropped area. These help in augmenting soil moisture and prevent erosion of topsoil. Contour bunds are effective methods to conserve soil moisture in watershed for long duration. Flowing water is intercepted before it attains the erosive velocity by keeping suitable spacing between bunds. Spacing between two contour bunds depends on the slope, the area and the permeability of the soil. Lesser the permeability of soil, the close should be spacing of bunds. This is also known as contour furrows and desert strip farming.

Gully plugging: These are soil and water retaining structures built across gullies in hilly areas. These are built with locally available materials like stone boulders, earth, brushwood, etc. The sites for gully plugs may be chosen whenever there is a local break in slope to permit accumulation of adequate water behind the bunds.

6.4.11 *In-situ* Water Conservation

Several methods like broad bed and furrow system, ridging and tied ridging, strip cropping, etc. may be practised for *in situ* conservation of water.

Broad bed and furrow system: The broad bed and furrow (BBF) system has been mainly developed at the International Crops Research Institute for the Semiarid Tropics (ICRISAT) in India. It is a modern version of the very old concept of encouraging controlled surface drainage by forming the soil surface into beds. The recommended ICRISAT system consists of broad beds about 100 cm wide separated by sunken furrows about 50 cm wide. The preferred slope along the furrow is between 0.4 and 0.8% on vertisols. Two, three, or four rows of crop can be grown on the broad bed, and the bed width and crop geometry can be varied to suit the cultivation and planting equipment. The major objectives are (i) to encourage moisture storage in the soil profile, (ii) to dispose safely of surplus surface runoff without causing erosion, (iii) to provide a better drained and more easily cultivated soil in the beds, and (iv) to reuse the runoff water stored in small tanks. The intercropping or sequential cropping is also possible on the broad-beds. Small amounts of life-saving irrigation applications can be very effective in dry spells during the rains, particularly on soils with lower storage capacity than the deep vertisols.

Ridging and tied ridging: This method is also known as furrow blocking, furrow damming and basin listing. The principle is to increase surface storage by first making ridges and furrows, then damming the furrows with small mounds or ties. Tied-ridging is a system of land configuration where semi-permanent ridges are prepared with crossties along the furrows to trap runoff of water. The ridges are generally laid across the main slope at a grade of 0.4-1%. Normally once constructed the ridges are not destroyed for a period of six seasons depending on the crop rotations practised by the farmer. Planting is done on top of the ridges. In subsequent seasons land preparation simply involves planting on top of the ridges without any tillage operation. For emergence and good establishment of the crop, planting is recommended only when the ridges are optimally moist. In drier areas planting may also be carried out in the furrows where most of the runoff water is collected.

Strip cropping: In strip cropping, the crop is sown in narrow, tilled rows along contours on a hillside. The strips of land between the rows, which are left untilled in natural grasses, slow the flow of rainwater down the slope and prevent it from washing away the topsoil. More water penetrates into the soil and provides moisture for the crop. The grass strips also provide a natural habitat for insects, many of which will prefer this to the crop. Organic matter and fertiliser are concentrated in furrows where the seed is planted. By tilling only the strips to be planted, the farmer saves labour in comparison with cultivation of the entire field. In subsequent seasons, the rows are replanted and the strips of grass cut back periodically, but they are never dug up.

6.5 CONSTRAINTS FOR ADOPTION OF LEISA

The following major constraints for adoption of LEISA techniques by the farmers have been identified by Kessler and Moolhuijzen (1994).

Reduced production and rehabilitation period

In most cases, introduction of a LEISA technique decreases agricultural production for a certain period. For example, when shifting from chemical fertilisers to organic manures it takes several years to build up the organic matter content and fauna of the soil for organic inputs to become effective. So in initial years the expected returns are not realised and this may strongly discourage the farmer to adopt the technology. Similarly, in case of alley cropping although the shrubs/trees occupy a good portion of agricultural land the return from the alley cropping is much delayed. The duration of the transition period varies depending upon the ecology of the site like soil, climate vegetation factors, etc.

Labour requirement

Most of the LEISA techniques are labour intensive. Labour shortage for small families and the migrant absentees is acute; particularly to collect huge quantity of organic manure and other related activities. Farmers' decisions to invest extra labour on LEISA techniques are greatly influenced by the expected benefits from it in relation to the availability of alternate source of income, may be from less risky off-farm activities.

Land ownership

The adoption of LEISA techniques is more where the farmers own their land and usufruct rights. Tenant farmers are discouraged and even prohibited going for the LEISA techniques.

Gender biasness

Women appear more rapid in adopting LEISA techniques. However, government organisations and non-government organisations are mainly staffed by men and their activities are oriented towards men. The constraints of LEISA, as for the agricultural development in general, affect more to women than men; women generally have less land ownership than men, and women have less control over the profits of their labour investments in LEISA techniques.

Partnership Technology Development approach

An important instrument to develop LEISA techniques is the Partnership Technology Development (PTD) approach by which site-specific and socioeconomically adapted techniques are developed and tested on-farm, based on the farmers' indigenous knowledge, skills and priorities, with support from the capacities of the external agents and agricultural science. However, in most cases, the LEISA techniques are introduced with the top-down way for getting quick results. PTD is time consuming, both for extensionists and farmers, and it requires a relatively high level training of the extensionists. But the extension agencies commonly lack sufficient dedicated and well-trained staff. On-farm demonstrations of LEISA techniques are rarely designed properly, nor cover for a sufficient long period to draw a firm conclusion. The potential of PTD to motivate the farmers is strongly related to the possibilities of obtaining short-term economic benefits.

Government policies

Often the policy-makers lack knowledge and appreciation of LEISA approaches. Government agencies provide subsidy and favourable credit system to the external inputs only for farm activities. There is no clear-cut policy to offer incentives to the farmers for adopting LEISA techniques or compensate the loss incurred during initial years of adoption.

Information dissemination

The availability of technical information and examples of successful implementation of LEISA techniques are important elements of PTD approach to motivate the farmers to adopt the LEISA techniques. However, this information is often unavailable.

7

Conservation Agriculture

Conservation agriculture is an approach to managing agroecosystems for improved and sustained productivity, increased profits and food security while preserving and enhancing the resource base and the environment. Food and Agriculture Organisation of the United Nations defines conservation agriculture as a farming system that promotes maintenance of a permanent soil cover, minimum soil disturbance (i.e. no tillage), and diversification of plant species.

7.1 CONCEPT OF CONSERVATION AGRICULTURE

Conservation agriculture enhances biodiversity and natural biological processes above and below the ground surface, which contribute to increased water and nutrient use efficiency and to improved and sustained crop production. Conservation agriculture is largely the product of the collective efforts of a number of previous agricultural movements, including no-till agriculture, agroforestry, green manures/cover crops (GMCC), direct planting/seeding, integrated pest management, and conservation tillage among many others. Yet conservation agriculture (CA) is distinct from each of these so called agricultural packages, even as it draws upon many of their core principles.

Thus, conservation agriculture is a set of soil management practices that minimise the disruption of the soil's structure, composition and natural biodiversity. Conservation agriculture is often confused with other terms including no-till/zero-till, conservation tillage/minimum tillage/reduced tillage, direct planting/direct drilling, direct seeding, organic farming. Though these terms are frequently used in CA practices, there is distinct difference among these terms.

No-till/ Zero till: No-till (NT) and zero till (ZT) are technical components used in conservation agriculture that simply involve the absence of tillage/ploughing operations on the soil. Crops are planted directly into a seedbed not tilled after harvesting the previous crop. It is not necessary that everyone utilising no-till technologies adopts other important components of CA. Another major difference is that NT or ZT do not necessarily leave residue mulch, which is an important practice in CA. Without the residue mulch many of the benefits of CA are lost or decreased.

Conservation tillage/ minimum tillage/ reduced tillage: These are tillage operations that leave at least 30% of the soil surface covered by plant residues in order to increase water infiltration and cut down on soil erosion and runoff. Conservation tillage is an intermediate form of CA since it keeps some soil cover as residue from the previous crop. But some tillage is usually done.

Direct planting/ direct drilling: In direct planting or direct drilling special equipment (e.g. NT drill) are used to plant seeds directly into crop residues left on the soil surface without preparing a seedbed beforehand.

Direct seeding: This term is usually associated with growing rice or any other cereal crop without producing seedlings that are then transplanted into the main field. However, the term direct seeding is used in CA systems as synonymous with no-till farming, zero tillage, no-tillage, direct drilling, etc.

Organic farming: Organic agriculture does not permit the use of synthetic chemicals to produce plant and animal products, relying instead on the management of soil organic matter and biological processes. But organic farming uses the principles of CA to some extent and one objective similar to CA is to maintain and improve soil health. Unlike organic farming, CA does allow farmers to apply synthetic chemical fertilisers, fungicides, pesticides and herbicides. These are used to control pests and weeds, particularly during the early transition years. As soil physical, chemical and biological health improves over time; the use of agrochemicals may be significantly reduced or, in some cases, phased out entirely.

Table 7.1. Some distinguishing features of conventional and conservation agriculture systems

Conventional agriculture	Conservation agriculture
Cultivating land, using science and technology to dominate nature	Least interference with natural processes
Excessive mechanical tillage and soil erosion	No-till or drastically reduced tillage (biological tillage)
High wind and soil erosion	Low wind and soil erosion
Residue burning or removal (bare surface)	Surface retention of residues (permanently covered)
Water infiltration is low	Infiltration rate of water is high
Use of *ex situ* FYM/composts	Use of *in situ* organics/composts
Green manuring (incorporated)	Brown manuring/cover crops (surface retention)
Kills established weeds but also stimulates more weed seeds to germinate	Weeds are a problem in the early stages of adoption but decrease with time
Free-wheeling of farm machinery, increased soil compaction	Controlled traffic, compaction in tramline, no compaction in crop area

Monocropping/monoculture, less efficient rotations	Diversified and more efficient rotations
Heavy reliance on manual labour, uncertainty of operations	Mechanised operations, ensure timeliness of operations
Poor adaptation to stresses, yield losses greater under stress conditions	More resilience to stresses, yield losses are less under stress conditions
Productivity gains in long-run are in declining order	Productivity gains in long-run are in incremental order

Source: Sharma et al., 2012

7.2 HISTORY OF CONSERVATION AGRICULTURE

Tillage, particularly in fragile ecosystems, was questioned for the first time in the 1930s, when the dustbowls devastated wide areas of the mid-west United States. Concepts for reducing tillage and keeping soil covered were introduced and the term conservation tillage was introduced to reflect such practices aimed at soil protection. Seeding machinery developments allowed then, in the 1940s, to seed directly without any soil tillage. At the same time theoretical concepts resembling today's conservation agriculture concept were developed by Edward Faulkner who published a book called 'Ploughman's Folly' in 1945 and Masanobu Fukuoka who published a book called 'One Straw Revolution' in 1975. But it was not until the 1960s for no-tillage to enter into farming practice in the USA. In the early 1970s no-till reached Brazil, where farmers together with extension scientists transformed the technology into the production system which today is called conservation agriculture. Yet it took another 20 years before CA reached significant adoption levels. During this time farm equipment and agronomic practices in no-till systems were improved and developed to optimise the performance of crops, machinery and field operations.

While tillage-based agriculture has been researched for several centuries, conservation agriculture is only half-a-century old and the functioning of CA systems can only be understood as the agroecosystems evolve under the new production management. From the early 1990s, the uptake of CA started growing exponentially, leading to a revolution in the agriculture of Brazil, Argentina, Paraguay and Uruguay. During the 1990s, this development increasingly attracted attention from farmers and researchers in Europe, Asia, Africa and Australia, and from development and international research organisations such as Food and Agriculture Organisation (FAO), World Bank, International Fund for Agricultural Development (IFAD), German Agency for International Cooperation (GIZ), Norwegian Agency for Development Cooperation (NORAD), French Agricultural Research Centre for International Development (CIRAD), Australian Centre for International Agricultural Research (ACIAR) and the

Consultative Group on International Agricultural Research (CGIAR) system. The improvement of conservation and no-tillage practices within an integrated farming concept such as conservation agriculture also led to a cropping system diversification and increased adoption of CA in industrialised countries after 2000. Conservation agriculture crop production systems are now becoming popular and experiencing increased interest worldwide. The total area under CA was 117 million hectares in 2010 (Kassam et al., 2010) and 180.44 million hectares in 2015 (Kassam et al., 2019), registering an increase of 54.2% in five years.

Table 7.2. Cropland area under conservation agriculture (M ha) in 2015-16

Region	CA cropland area of the region	Percent of global CA cropland area	Percent of cropland area in the region
South America	69.90	38.7	63.2
North America	63.18	35.0	28.1
Australia & New Zealand	22.67	12.6	45.5
Asia	13.93	7.7	4.1
Russia & Ukraine	5.70	3.2	3.6
Europe	3.56	2.0	5.0
Africa	1.51	0.8	1.11
Global total	180.44	100	12.5

Source: Kassam *et al.*, 2019

7.3 OBJECTIVES OF CONSERVATION AGRICULTURE

The conservation agriculture offers a range of productivity, socioeconomic and environmental benefits to producers and to society at large on a sustainable basis. These are based on five overall objectives.

1. Simultaneous achievement of increased agricultural productivity and enhanced ecosystem services
2. Enhanced input-use efficiency, including water, nutrients, pesticides, energy, land and labour
3. Judicious use of external inputs derived from fossil fuels (such as mineral fertilisers and pesticides) and preference for alternatives (such as recycled organic matter, biological nitrogen fixation and integrated pest management)
4. Protection of soil, water and biodiversity through use of 'minimum soil disturbance' and maintaining organic matter cover on the soil surface to protect the soil and enhance soil organic matter and soil biodiversity
5. Use of managed and natural biodiversity of species to build systems' resilience to abiotic, biotic and economic stresses, with an underlying emphasis on

improving soils' content of organic matter as a substrate essential for the activity of the soil biota

The farming practices required to implement these objectives will differ according to local conditions and needs, but will have the following management practices in common.

1. Minimising soil disturbance by mechanical tillage and, whenever possible, seeding or planting directly into untilled soil, in order to maintain soil organic matter, soil structure and overall soil health
2. Enhancing and maintaining organic matter cover on the soil surface, using crops, cover crops or crop residues
3. Diversification of species, both annuals and perennials, in associations, sequences and rotations that can include trees, shrubs, pastures and crops, all contributing to enhanced crop nutrition and improved system resilience
4. Use of well adapted, high yielding varieties and good quality seeds
5. Enhanced crop nutrition, based on healthy soils
6. Integrated management of pests, diseases and weeds
7. Efficient water management

Sustainable crop production intensification is the combination of all seven of these improved practices applied in a timely and efficient manner. Such sustainable production systems are knowledge- and management-intensive and relatively complex to learn and implement. They offer farmers many possible combinations of practices to choose from and adapt, according to their local production conditions and constraints (Kassam et al., 2018).

7.4 PRINCIPLES OF CONSERVATION AGRICULTURE

Conservation agriculture practices perused in many parts of the world are built on ecological principles making land use more sustainable (Lal, 2013). Adoption of CA for enhancing resource use efficiency and crop productivity is the need of the hour as a powerful tool for management of natural resources and to achieve sustainability in agriculture (Bhan and Behera, 2014). The FAO has determined that CA has three key principles that farmers can proceed through in the process of conservation agriculture. These principles are linked and must be considered together for appropriate design, planning and implementation processes.

These three principles are

1. Minimum mechanical soil disturbance (i.e. no tillage) through direct seed and/or fertiliser placement

2. Permanent soil organic cover (at least 30%) with crop residues and/or cover crops
3. Species diversification through varied crop sequences and associations involving at least three different crops

Conservation agriculture principles are universally applicable to all agricultural landscapes and land uses with locally adapted practices. Conservation agriculture enhances biodiversity and natural biological processes above and below the ground surface. Soil interventions such as mechanical tillage are reduced to an absolute minimum or avoided, and external inputs such as agrochemicals and plant nutrients of mineral or organic origin are applied optimally and in ways and quantities that do not interfere with, or disrupt, the biological processes. Conservation agriculture facilitates good agronomy, such as timely operations, and improves overall land husbandry for rainfed and irrigated production. Complemented by other known good practices, including the use of quality seeds, and integrated pest, nutrient, weed and water management, etc., CA is a base for sustainable agricultural production intensification. It opens increased options for integration of production sectors, such as crop-livestock integration and the integration of trees and pastures into agricultural landscapes (FAO, 2010).

Conservation agriculture can contribute to sustainable agriculture and rural development by improving input use efficiency, increasing farm income, sustaining or increasing agricultural land, and protecting the natural resource base (FAO, 2007).

7.4.1 Minimum Tillage and Soil Disturbance

The first key principle in CA is practising minimum soil disturbance which is essential for maintaining minerals within the soil, stopping erosion, and preventing water loss from the soil. In the conventional agriculture, soil tillage is considered as a main process in the introduction of new crops to an area. It is believed that tilling the soil would increase fertility within the soil through mineralisation that takes place in the soil. However, in conservation agriculture, tillage is considered as a harmful practice that destroys organic matter that is found within the soil cover and causes severe erosion and crusting which leads to a decrease in soil fertility.

The soil biological activity produces very stable soil aggregates as well as various sizes of pores, allowing air and water infiltration. This process can be called 'biological tillage' and it is not compatible with mechanical tillage. With mechanical soil disturbance, the biological soil structuring processes will disappear. Minimum soil disturbance provides/maintains optimum proportions of respiration gases in the rooting-zone, moderate organic matter oxidation, porosity for water movement, retention and release and limits the re-exposure of weed seeds and their germination.

Direct seeding involves growing crops without mechanical seedbed preparation and with minimal soil disturbance since the harvest of the previous crop. Planting refers to the precise placing of large seeds (for example, maize and beans); whereas seeding usually refers to a continuous flow of seed as in the case of small cereals (for example, wheat and barley). The equipment penetrates the soil cover, opens a seeding slot and places the seed into that slot. The size of the seed slot and the associated movement of soil are to be kept at the absolute minimum possible. Ideally the seed slot is completely covered by mulch again after seeding and no loose soil should be visible on the surface.

No-till farming is a process that can save soil organic levels for a longer period and still allow the soil to be productive for longer periods (FAO, 2007). Additionally, the process of tilling can increase time and labour for producing that crop. Minimum soil disturbance also reduces destruction of soil micro- and macro-organism habitats that is common in conventional ploughing practices.

Land preparation for seeding or planting under no-tillage involves slashing or rolling the weeds, previous crop residues or cover crops; or spraying herbicides for weed control, and seeding directly through the mulch. Crop residues are retained either completely or to a suitable amount to guarantee the complete soil cover, and fertiliser and amendments are either broadcast on the soil surface or applied during seeding.

7.4.2 Permanent Soil Organic Cover

Keeping the soil covered is a fundamental principle of CA. A permanent soil cover is important to protect the soil against the deleterious effects of exposure to rain and sun; to provide the micro- and macro- organisms in the soil with a constant supply of 'food'; and alter the microclimate in the soil for optimal growth and development of soil organisms, including plant roots. In turn it improves soil aggregation, soil biological activity and soil biodiversity and carbon sequestration. Crop residues are left on the soil surface, but cover crops may be required if the gap is too long between harvesting one crop and establishing the next. Cover crops improve the stability of the CA system, not only on the improvement of soil properties but also for their capacity to promote an increased biodiversity in the agroecosystem. The principle of managing the top soil to create a permanent organic soil cover can allow for growth of organisms within the soil structure. This growth will break down the mulch that is left on the soil surface. The breaking down of this mulch will produce a high organic matter level which will act as a fertiliser for the soil surface.

If conservation agriculture is in practice for many years and enough organic matter was being built up at the surface, then a layer of mulch would start to form. This layer helps prevent soil erosion from taking place and ruining the soil's profile. The presence of mulching also reduces the velocity of runoff and the impact of raindrops thus reducing soil erosion and runoff. The layer of mulch that is built up over time will become like a buffer zone between soil and mulch and this will help reduce wind and water erosion. This type of ground cover also helps keep the temperature and moisture levels of the soil at a higher level rather than if it was tilled every year.

The presence of a mulch layer (of dead vegetation) in conservation agriculture inhibits the evaporation of soil moisture, yet leads to greater water infiltration into the soil profile. The percentage of rainwater that infiltrates the soil depends on the amount of soil cover provided. Vegetative cover is important in CA for the protection of the soil against the impacts of raindrops; to keep the soil shaded; and maintain the highest possible moisture content. Straw residues function as a cushion that reduces the pressure on the soil under wheels and hooves and so they play an important role in reducing soil compaction.

While commercial crops have a market value, cover crops are mainly grown for their effect on soil fertility or as livestock fodder. Cover crops are grown during fallow periods, between harvest and planting of commercial crops, utilising the residual soil moisture. Their growth is interrupted either before the next crop is sown, or after sowing the next crop, but before competition between the two crops starts. In regions where smaller amounts of biomass are produced, such as semiarid regions or areas of eroded and degraded soils, cover crops are very useful.

The cover crops provide the following functions.

- Protect the soil during fallow periods.
- Recycle nutrients (especially P and K) and mobilise them in the soil profile in order to make them more readily available to the following crops.
- Improve the soil structure and break compacted layers and hard pans.
- Permit a rotation in a monoculture.
- Can be used to control weeds and pests.
- Provide an additional source of organic matter to improve soil structure.
- Provide 'biological tillage' of the soil; the roots of some crops, especially cruciferous crops are able to penetrate compacted or very dense layers. increasing water percolation capacity of the soil.
- Utilise easily leached nutrients, especially N.

7.4.3 Species Diversification

The third principle is the practising diverse crop rotations or crop interactions. The rotation of crops is not only necessary to offer a diverse 'diet' to the soil microorganisms, but as they root at different soil depths, they are capable of exploring different soil layers for nutrients. Nutrients that have been leached to deeper layers and that are no longer available for the commercial crop can be 'recycled' by the crops in rotation. This way the rotation crops function as biological pumps. Furthermore, a diversity of crops in rotation leads to a diverse soil flora and fauna, as the roots excrete different organic substances that attract different types of bacteria and fungi, which in turn, play an important role in the transformation of these substances into plant available nutrients. Cropping sequence and rotations involving legumes helps in minimal rates of build-up of population of pest species, through life cycle disruption, biological nitrogen fixation, control of off-site pollution and enhancing biodiversity.

Crop rotation also has an important phytosanitary function as it prevents the carryover of crop-specific pests and diseases from one crop to the next via crop residues. Rotational crops will act as a natural insecticide and herbicide against specific crops. Not allowing insects or weeds to establish a pattern will help to eliminate problems with yield reduction and infestations within fields (FAO, 2007). Crop rotation can also help build-up soil infrastructure. Establishing crops in a rotation allows for an extensive build-up of rooting zones which will allow for better water infiltration. Organic molecules in the soil break down into phosphates, nitrates and other beneficial elements which are thus better absorbed by plants. Ploughing increases the amount of oxygen in the soil and increases the aerobic processes, hastening the breakdown of organic material. Thus, more nutrients are available for the next crop but, at the same time, the soil is depleted more quickly of its nutrient reserves.

The effects of crop rotation can be summarised as follows.

- Higher diversity in plant production and thus in human and livestock nutrition
- Reduction and reduced risk of pest and weed infestations
- Greater distribution of channels or biopores created by diverse roots (various forms, sizes and depths)
- Better distribution of water and nutrients through the soil profile
- Exploration for nutrients and water of diverse strata of the soil profile by roots of many different plant species resulting in a greater use of the available nutrients and water

- Increased nitrogen fixation through certain plant-soil biota symbionts and improved balance of N/P/K from both organic and mineral sources
- Increased humus formation

7.5 ADVANTAGES OF CONSERVATION AGRICULTURE

Conservation agriculture requires a new way of thinking about agricultural production in order to understand how one could possibly attain higher yields with less labour, less water and fewer chemical inputs. In spite of these challenges, conservation agriculture is spreading to farmers throughout the world as its benefits become more widely recognised by farmers, researchers, scientists and extensionists alike.

7.5.1 Economic Considerations

Farmers using conservation agriculture technologies typically report higher yields with less water, fertiliser and labour inputs, thereby resulting in higher overall farm profits. CA increases the productivity of land, labour, water, nutrients and soil biota.

Land: Conservation agriculture improves soil structure and protects the soil against erosion and nutrient losses by maintaining a permanent soil cover and minimising soil disturbance. Furthermore, CA practices enhance soil organic matter levels and nutrient availability by utilising the previous crop residues or growing green manure/ cover crops and keeping these residues as surface mulch rather than burning. Thus, arable land under CA is more productive for much longer periods of time.

Labour: Much of the reduced labour comes from the absence of tillage operations under CA, which use up valuable labour days during the planting season. The labour and time saving under CA enables farm families to invest their time in more rewarding activities such as processing of raw produce or other value adding activities, enhancing their farm income than the crop production would do.

Water: Conservation agriculture requires significantly less water use due to increased infiltration and enhanced water holding capacity from crop residues left on the soil surface. Mulches also protect the soil surface from extreme temperatures and greatly reduce surface evaporation, which is particularly important in tropical and subtropical climates. CA reduces the vulnerability of crops against unexpected extreme climatic events. It reduces crop water requirements by 30%, making better use of soil water and facilitating deeper rooting of crops, so that drought periods can be tolerated better.

Nutrients: Soil nutrient supplies and cycling are enhanced by the biochemical decomposition of organic crop residues at the soil surface that are also vital for feeding the soil microbes. While much of the nitrogen needs of primary food crops can be achieved by planting nitrogen-fixing legume species, other plant essential nutrients often must be supplemented by additional chemical and/or organic fertiliser inputs. In general, soil fertility is built up over time under conservation agriculture, and fewer fertiliser amendments are required to achieve optimal yields over time. Legumes in CA rotations provide increased *in situ* availability of nitrogen, thus diminishing the need for large amounts of applied nitrogenous fertilisers.

Soil biota: Insect pests and other disease causing organisms are held in check by an abundant and diverse community of beneficial soil organisms, including predatory wasps, spiders, nematodes, springtails, mites and beneficial bacteria and fungi, among other species. Furthermore, the burrowing activity of earthworms and other fauna create tiny channels or pores in the soil that facilitate the exchange of water and gases and loosen the soil for enhanced root penetration.

Energy savings: The benefits of lower costs for fuel and/or draft animals are straightforward. Farmers who use tractors to plough are able to reduce their fuel use for ploughing by two-thirds. Enhanced soil structure makes soil more easily trafficked by farm machinery for the remaining operations, further reducing fuel costs, as well as equipment wear and tear.

Time savings: Time savings allow farmers to plant earlier. Depending on the agricultural environment, this can improve the odds that the crop will mature and be ready to harvest before the onset of late-season drought. It may allow farmers to squeeze in a third cropping per year in what had been a system producing only two crops per year, adding perhaps a crop of high-value vegetables or another cash crop.

7.5.2 Environmental Considerations

Intensive agriculture is often considered to be a damaging factor to ecosystems in view of the negative impact it has on rural landscape, soil resources, water quality and biodiversity. The impact of agriculture on ecosystems through erosion, pollution of waterbodies and dust and smoke emissions is also felt outside the actual agricultural area. However, conservation agriculture represents an environment-friendly set of technologies. Because it uses resources more efficiently than conventional agriculture, these resources become available for other uses, including conserving them for future generations. The significant reduction in fossil fuel use under no-till agriculture results in fewer greenhouse gases being emitted into the atmosphere and cleaner air in general. A reduced application of agrochemicals under CA also significantly lessens pollution levels in air, soil and water. Society gains from CA, regardless of farm size by diminished erosion and runoff, less downstream sedimentation and flood damage to infrastructure, better recharge of groundwater, more regular stream flow throughout the year, resulting in a more assured community water supply and quality.

While the economic advantages of conservation agriculture are often realised with the first cropping, the environmental advantages emerge more slowly. Most environmental advantages require majority adoption in a community, subsequently maintained year after year, to avoid being negated by environmental pollution flowing from conventional farms.

Natural resource base: Conservation agriculture reverses soil degradation processes and builds up soil fertility and productive capacity. It facilitates a better infiltration of rainwater, enabling the recharge of groundwater resources while at the same time reducing the pollution of waterbodies through reduced erosion and leaching. Conservation agriculture conserves and enhances natural resources while maintaining and sustainably increasing production levels.

Air: A rising problem across regions is rural air pollution caused by farmers burning crop stubble and other residues. If conservation agriculture becomes prevalent enough in a farm area, leaving most residues in the field for soil protection, fires and the haze they cause can become a thing of the past. Undisturbed soil is less prone to rise as dust, which is another rural health hazard when ploughed soil is left bare between crops.

Surface water: Improved water infiltration into healthy soil and reduced water erosion lessens the nuisance and threats to public and environmental health that occur when agricultural runoff causes excessive silting of rivers, reservoirs, lakes and micro catchments. As conservation agriculture eliminates soil erosion and greatly reduces runoff, widespread adoption in a farm area can be expected to enhance the cleanliness and biodiversity of surface waterbodies, protecting their value as fisheries, sources of drinking water and venues for recreation.

Landscape: The avoidance of ploughing in CA facilitates the introduction of trees and hedgerows into the agricultural landscape in a closer vicinity of field crops than under tillage based agriculture. The greater diversity in the crop rotations also contributes to a more diverse and pest-free landscape. It also increases biodiversity in the agricultural production systems.

Climate change: Conservation agriculture can contribute to reduced greenhouse gas emissions from agricultural crop production through reduced fuel use, better aeration of soils that reduces nitrous oxide emissions and, in no-till non-flooded rice, methane emissions. In addition it binds atmospheric carbon in the soil in the form of soil organic matter. With this, CA helps to mitigate climate change.

Under CA systems, it is possible to harness many of the above ecosystem services mainly because the ecosystem functions that generate these services are enhanced and protected, so that agriculture is not in competition with nature but works in harmony with it. However, the environmental advantages of conservation agriculture emerge more slowly than the economic advantages which are often realised with the first cropping. Most environmental advantages require majority adoption in a community, subsequently maintained year after year, to avoid being negated by environmental pollution flowing from conventional farms.

7.5.3 Sociocultural Considerations

The rural community ultimately benefits from better nutrition, overall health and less pressure on curative health services.

Equity considerations: Conservation agriculture has the benefit of being accessible to many small-scale farmers who need to obtain the highest possible yields with limited land area and inputs. Perhaps the biggest obstacle for the technology spreading to more small-scale farmers worldwide has been limited access in certain areas to certain specialised equipment and machinery, such as no-till planters. This problem can be remedied by available service providers renting equipment or undertaking conservation agriculture operations for farmers who would not otherwise have access to the needed equipment. Formulating policies that promote adoption of CA are also needed. As more and more small-farmers gain access to CA technologies, the system becomes much more 'scale neutral'.

Active role for farmers: As with any new agricultural technology, CA methods are most effective when used with skilful management and careful consideration of the many agroecological factors affecting production on any given farm or field. Rather than being a fixed technology to be adopted in blueprint-like fashion, CA should be seen as a set of sound agricultural principles and practices that can be applied either individually or together, based on resource availability and other factors. For this reason, farmers are encouraged to experiment with the methods and to evaluate the results for themselves, not just to 'adopt' CA technologies. Selecting among different cover crop species, for example, needs to be determined in relation to particular agroecological conditions of the farm, including soil type, climate, topography as well as seed availability and what the primary function of the GMCC will be. Similarly, planting distances, irrigation requirements and the use of agrochemicals to control weeds and pests among other considerations, must be decided based on what the farmer needs as well as the availability of these and other resources.

7.6 CONSTRAINTS FOR ADOPTION OF CONSERVATION AGRICULTURE

Socioeconomic and psychological factors are significant in the decision making process to adopt the CA. Farmers who have a strong conservation ethic, for example, may be willing to accept reduced profits in return for feeling that they have contributed to welfare of future generations. Environmentally concerned farmers may also be willing to invest in practices that will enhance the environmental quality of their lands and water resources, as well as enhance the economic value of land when it is passed on to next generation.

The most important factor in the adoption of CA is overcoming the bias or mindset about tillage. Convincing the farmers that successful cultivation is possible even with reduced tillage or without tillage is a major hurdle in promoting CA on a large scale. CA is now, considered a route to sustainable agriculture. The constraints which hampered the adoption of CA have been identified by various researchers (Meena and Singh, 2013) which can be listed under four broad categories; technical, extension, financial and production. These prevailing constraints strictly control the adoption behaviour at farmers' level.

Technical constraints

- Lack of appropriate seeders especially for small and marginal farmers
- Non-availability of quality drill
- Lack of regular monitoring of machines
- Lack of training/ capacity building
- Non-availability of spare parts in local market
- Lack of local manufacturers of machines

Extension constraints

- Lack of knowledge about the potential of CA to agriculture leaders and farmers
- Lack of extension support from state extension agencies
- Lack of extension literature
- Lack of attention by mass media
- Lack of knowledge of extension agencies

- Inadequate extension facility at disposal of input agencies
- Lack of cooperation from fellow farmers

Financial constraints

- Lack of credit facilities
- Lack of money to buy new machines and inputs
- No subsidy on machines
- High cost of drill

Production constraints

- Wide spread use of crop residues for livestock feed and fuel
- Scarcity of crop residues due to less biomass production of crops under rainfed situations
- Burning of crop residues by the farmers for timely sowing of the next crop
- Non-availability of skilled and scientific manpower
- Free grazing of animals that compact the soil and remove all the soil cover, leaving it open to erosion
- Difficulty in getting seeds of cover crops and herbicides

7.7 CHALLENGES IN CONSERVATION AGRICULTURE

Understanding the system

Conservation agriculture systems are much more complex than conventional systems. Site specific knowledge has been the main limitation to the spread of CA system. Managing these systems efficiently will be highly demanding in terms of understanding of basic processes and component interactions, which determine the whole system performance. Thus, the need is to recognise conservation agriculture as a system and develop management strategies.

Building a system perspective

A system perspective is built working in partnership with farmers. A core group of scientists, farmers, extension workers and other stakeholders working in partnership mode will therefore be critical in developing and promoting new technologies. This is somewhat different than in conventional agricultural R&D, the system is to set research priorities and allocate resources within a framework, and little attention is given to build relationships and seek linkages with partners working in complementary fields.

Technological challenges

While the basic principles which form the foundation of conservation agriculture practices, that is, no tillage and surface managed crop residues are well understood, adoption of these practices under varying farming situations is the key challenge. These challenges relate to development, standardisation and adoption of farm machinery for seeding with minimum soil disturbance, developing crop harvesting and management systems.

Site specificity

Adapting strategies for conservation agriculture systems will be highly site specific, yet learning across the sites will be a powerful way in understanding why certain technologies or practices are effective in a set of situations and not effective in another set. This learning process will accelerate building a knowledge base for sustainable resource management.

Long-term research perspective

Conservation agriculture practices, e.g. no-tillage and surface maintained crop residues result in resource improvement only gradually, and benefits come about only with time. In conservation agriculture, benefits in terms of yield increase may not come in the early years of practices. Understanding the dynamics of changes and interactions among physical, chemical and biological processes is basic to developing improved soil-water and nutrient management strategies. Therefore, research in conservation agriculture must have longer term perspectives and should focus on several innovative features to address the challenge.

7.8 STRATEGY FOR IMPLEMENTATION OF CONSERVATION AGRICULTURE

After evaluating determinants and constraints in adoption of CA, it is the prime function of extension workers to diffuse new ideas and practices among farmers. It is their task to expedite the process of getting ideas from their sources of origin to those who can adopt or use them. The following strategies have been suggested by Meena and Singh (2012).

Implementing situations and prevailing constraints

Factors limiting the agricultural production should be rectified before the full benefits from implementation of CA can be realised. This might refer to technical factors, such as soil compaction, insufficient drainage, soil chemical properties, as well as socioeconomic factors such as availability of adequate technology, investment capital, land use rights, livestock pressure, customary practices or access to markets. These will have to be addressed in order to establish CA in a sustainable manner.

Transforming the agricultural system

The transition phase usually takes about two years; however, the full benefits of this system often become visible only after five years. In CA, mechanical tillage is replaced by biological tillage and soil fertility is essentially managed through soil cover management, crop rotations and weed management. Fertilisers, water harvesting technologies and irrigation can complement CA, and minimum tillage might be necessary in some cases particularly during the transition.

Changing the attitude

Changing the attitude or mentality of the farming community is a difficult task but it paves the way to success for task like implementing the conservation agriculture. Proper knowledge about the concepts of conservation agriculture is also inevitable like soil is a habitat for roots and soil organisms, any damage to this habitat endangers soil fertility and leads to land degradation, soil fauna creates a stable soil structure, etc.

Encouragement, support and capacity building

Promotion of CA should be done simultaneously through policies, education, research and extension institutions in the field. Adoption by farmers is supported most effectively through farmers' groups, study tours, networks and NGOs. Research and extension institutions and the private sector have a major role in providing farmers with appropriate and affordable technologies.

Policies and incentives

Conservation agriculture implies a radical change from traditional agriculture. There is need for policy analysis to understand how CA technologies integrate with other technologies, and how policy instruments and institutional arrangements promote or deter CA. Thus, policies should focus on access to market, credit and input supplies, and rural infrastructures. Policies should support the development of farmers' groups. Incentives should encourage diversification and CA practices, especially during the transition phase. Inadequate policies and subsidies that support conventional practices might constrain CA adoption. Land use and customary rights must also be taken into account and eventually adapted to favour CA adoption by farmers and rural communities.

Support from international organisations

Food and Agricultural Organisation is promoting the adoption of the CA concept at policy level as well as stimulating farmer-based movements and collaboration between the research sector and farmer groups. Due to its positive effects on food security, biodiversity, land and water resources, carbon sequestration and sustainable development, CA is a major opportunity to implement the international conventions on combating desertification, on biodiversity and on climate change.

7.9 OPERATIONAL FACTORS FOR SUCCESS OF CONSERVATION AGRICULTURE

Maintaining permanent or semi-permanent soil cover

Plants are left growing or killed and their residues left to decompose *in situ*. The primary function of this is to protect organic matter-enriched topsoil against chemical and physical weathering. Plant residues intercept energy from falling raindrops, provide a barrier from strong winds, and moderate temperatures, improving water infiltration and decreasing surface evaporation from sunlight. Surface cover also favours enhanced levels of biological activity by providing food for soil microbes, especially in tropical and subtropical areas.

Minimum soil disturbance

No-till does not involve any loosening of the soil except for a very small area immediately surrounding where the seed is planted. This lack of soil disturbance serves to maintain overall soil structure, including aggregate stability and porosity, both of which promote the exchange of water and gases and provide habitat to an abundant and diverse population of soil biota.

Regular crop rotations

Well-balanced crop rotations can neutralise many of the pest and disease problems associated with not tilling the soil, including the proliferation of insect pests and other harmful bacteria, viruses and fungi, by increasing the diversity and abundance of beneficial soil biota that can help keep pest and disease problems in check. Rotating crops also interrupts the life cycle of many weeds, thereby leading to a reduction in overall weed growth.

Utilisation of green manures/cover crops

Cover crops are grown specifically to help maintain soil fertility and productivity by decreasing erosion and/or by adding fresh plant residues to the soil. Leguminous cover crops offer the added advantage of being able to fix nitrogen from the atmosphere and add it to the soil, thereby increasing overall nitrogen availability for other crops. Cover crops are usually mowed, sprayed with chemical herbicides or otherwise killed before or during soil preparation for the next economic crop. It is generally recommended to leave a week or two between the killing of the cover crop and the planting of a primary crop in order to allow for some decomposition to occur as well as to lessen the effects of nitrogen immobilisation and allelopathic effects.

No burning of crop residues

Since crop residues are the principal element of permanent soil cover, they must never be burned or otherwise removed from the soil surface. Rather, plant residues are left on the soil surface in order to protect organic matter enriched topsoil from erosion while also adding fresh organic matter upon decomposition. Burning not only creates significant air pollution but also dramatically increases mineralisation rates, leading to the rapid depletion of soil organic matter and nutrients from the soil. However, in some situations farmers need to think of the trade-off between removing residues to feed their animals and leaving them to feed the soil. A win-win situation would do both and as yields and biomass increase over time, both become more feasible.

Integrated disease and pest management

Conservation agriculture depends heavily on enhanced biological activity to help control insect pests and other disease causing soil organisms. Integrated pest management entails the judicious use of crop rotations and other beneficial plant associations as well as chemical pesticides, herbicides and fungicides to control insect pest and disease problems. Over time, the enhanced biological activity and abundance brought on by no-till and other CA technologies results in decreased applications of agrochemicals.

Reduction in fossil fuel use and greenhouse gas emissions

Compared to conventional tillage, which often requires 4-8 tractor passes in a typical growing season, no-till significantly reduces the use of tractors and other heavy farm machinery and thus diesel. Furthermore, the increased levels of soil organic matter and plant-available nitrogen typically found in CA soils greatly reduces the need for applying large amounts of chemical fertilisers, many of which require significant fossil fuel energy to process. Thus, overall fossil fuel use and greenhouse gas emissions are greatly reduced.

Controlled / limited human and mechanical traffic over agricultural soils

The number of tractor passes over a given field is significantly reduced under CA, as compared to conventional tillage systems. However, increased bulk densities have been reported under CA, though this can be corrected by limiting the use of heavy farm machinery when soils are wet and most prone to compaction and/or by converting to a permanent raised bed system.

Timely availability of inputs and equipment

The farmers should have access to the right equipment. Timely availability of right seeds for the cover crops should be ascertained. Other inputs like fertiliser and/or manure, herbicides, etc. should be available timely. Conservation agriculture generally saves work. However, it may require more work in the first year.

Timely field operations

A key principle of conservation agriculture is the need for timely implementation. The land should be prepared well before the rain starts and planting is carried out soon after an effective rainfall event. Weeding should be at appropriate times and intervals. The pests and diseases should be effectively controlled before their spread too widely.

Precise operations

Precise measurements of row and plant spacing, evenness of depth and placement of soil amendments and covering of seed are also important. Planting should be done on the same lines each season. A key benefit is that compaction of the soil by feet, hooves and wheels will then only occur in the inter-row spaces and not over the crop lines. Residual fertility builds up in the rows and the crop roots of each consecutive crop provide organic matter.

Efficient use of inputs

Precise application of soil amendments leads to very little wastage because only the crops get benefitted, and not the surrounding soil and weeds. This gives higher yields and huge savings on costly inputs. Inputs like time, energy, draught power, etc. are used more effectively under conservation agriculture. Since land preparation can start soon after harvesting up until the rains start, labour inputs can be spread out more evenly over the year.

7.10 POLICY AND INSTITUTIONAL SUPPORT FOR CONSERVATION AGRICULTURE

In India and elsewhere much research work on conservation agriculture has been conducted. However, its percolation to farmers is very limited. So there is a need to think about the problems faced at the implementing level and devise a strategy involving all who are concerned. One of the reasons for poor percolation of the technology to the farmers is the mindset about tillage by the majority of farmers. Even the success rate is reported to be at low level where CA is practised. FAO (2001) has reported that this is partly because policy environments are not favourable. Under such situations, farmers' participatory on-farm research to evaluate/refine the technology in initial years followed by large scale demonstration in subsequent years is needed. In India, efforts are being initiated through a network research project for on-farm evaluation and demonstration of CA technology for its promotion. The following are some of the important policy considerations identified by Bhan and Behera (2014) for promotion of conservation agriculture.

Shift in focus from food security to livelihood security

The 'food security' policy based on cereal production must now replace a well-articulated policy goal for livelihood security. This will help the diversification of dominant rice-wheat cropping systems in the Indo-Gangetic Plains, the cultivation of which in conventional tillage practice has overexploited the natural resources in the region. The nature of cropping patterns and the extent of crop diversification are influenced by policy interventions. The government policies that directly or indirectly affect crop diversification include pricing policy, tax and tariff policies, trade policies and policies on public expenditure and agrarian reforms (Behera et al., 2007).

Development of CA equipments for diversified cropping systems

The success of conservation agriculture depends upon the development and standardisation of equipments for seeding, fertiliser placement and harvesting with minimum soil disturbance in residue management for different edaphic conditions. Thus, ensuring quality and availability of equipment through appropriate incentives will be important. In these situations, the subsidy support from national or local government to firms for developing low cost machines will help in the promotion of CA technologies.

Changes in the plant growing micro-environment due to CA technologies

These include changes in moisture regimes, root environment, emergence of new pathogens and shift in insect-pest scenario, etc. The requirement of plant types suited to the new environment, and to meet specific mechanisation-needs could be different. Therefore, there is a need to develop complementary crop improvement programmes, aimed at developing cultivars which are better suitable to new systems. Farmers' participatory research would appear promising for identifying and developing crop varieties suiting to a particular environment or locations.

Policy support for capacity building by organising training on CA

Availability of trained human resources at ground level is one of the major limiting factors in adoption of CA. Efforts to adequately train all new and existing agricultural extension personnel on CA should be made in relevant departments. In the long-term, CA should be included in curricula from primary school to university levels, including agricultural colleges. Inclusion of conservation and sustainability concepts in the course curricula with a suitable blend of biophysical and social sciences would be important for sustainable resource management.

Institutionalisation of CA

Institutionalising CA into relevant government ministries and departments and regional institutions is required for sustainability of the technology. Local, national and regional policy and decision makers could spearhead and support the formulation and development of strategies and mechanisms for scaling up the technology. CA could be integrated into interventions such as seed, fertiliser and tillage and draft power support programmes as a way of further enhancing productivity.

Support for the adaptation and validation of CA technologies in local environments

Adaptive research is required to tailor CA principles and practices to local conditions. This should be done in collaboration with local communities and other stakeholders. Issues that should be addressed include crop species, selection and management of crop and cover crop and rotations, maintenance of soil cover and CA equipment. The resource poor and small holder farmers do not have economic access to new seeds, herbicides and seeding machineries, etc. This calls for policy frame work to make easily available critical inputs.

Generation of resource database and linkages

There is a need for generating a good resource database on conservation agriculture practices and technologies. The agencies involved in conservation agriculture R&D should complement each others' work. Much attention should be given to build relationships and seek linkages with partners working in the complementary fields. Besides resources, systematic monitoring of the socioeconomic, environmental and institutional changes should become an integral part of the major projects on CA.

Rewards for environmental services and fines for faulty practices

Adopters of CA improve the environment through carbon sequestration, prevention of soil erosion or the encouragement of groundwater recharge. It provides ecosystem services, thus, farmers could be rewarded for such services, which have a great impact on the quality of life for all. Similarly, there should be a strong policy intervention for prohibiting unscientific and environment polluting practice like residue burning by imposing a fine.

Building partnership

Conservation agriculture systems are very complex and their efficient management needs understanding of basic processes and component interactions which determine the system performance. A system perspective is the best to build working in partnership with farmers, who are at the core of farming systems. Scientists, farmers, extension agents, policy makers and other stakeholders in the private sector working in partnership mode will be important in developing and promoting new technologies. Instead of using a top-down approach where the extension agent places CA demonstrations in farmer fields and expects the farmer to adopt, a more participatory system is required where the farmers are enabled through provision of equipment and training to experiment with the technology and find out for themselves whether it works and what fine-tuning is needed to make it successful on their land.

Credit and subsidy

There is a need to provide credit to farmers to buy the equipment, machinery, and inputs through banks and credit agencies at reasonable interest rates. The government should also provide a subsidy for the purchase of such equipment by farmers.

8

Integrated Farming Systems

Integrated farming system is a multidisciplinary whole farm approach, which enables the farmers to identify opportunities and threats and act accordingly. It is a dynamic approach which can be applied to any farming system around the world. It is very effective in solving the problems of small and marginal farmers. The approach aims at increasing income and employment from small-holding by integrating various farm enterprises and recycling crop residues and by-products within the farm itself. It involves attention to detail and continuous improvement in all areas of a farming business through informed management processes. Integrated farming combines the best of modern tools and technologies with traditional practices available at a given location and situation. Preserving and enhancing soil fertility, maintaining and improving a diverse environment and the adherence to ethical and social criteria are indispensable basic elements of integrated farming systems.

8.1 CONCEPTS AND DEFINITIONS OF IFS

The International Organisation of Biological Control describes integrated farming as a farming system where high quality food, feed, fibre and renewable energy are produced by using resources such as soil, water, air and nature as well as regulating factors to farm sustainably and with as little polluting inputs as possible. Integrated farming system (IFS) is a commonly and broadly used word to explain a more integrated approach to farming as compared to monoculture approaches. In this system an interrelated set of enterprises are maintained and by-products or wastes from one production system becomes an input for another production system, which reduces cost and improves production and/or income. Thus, IFS works as a system of systems. FAO (1977) stated that 'there is no waste', and 'waste is only a misplaced resource which can become a valuable material for another product' in IFS. For example, paddy straw, by-product from rice crop can be used as a valuable input for mushroom cultivation or dry fodder for dairy animals. Similarly spent of mushroom cultivation (used straw) can be used as a raw material in compost or vermicompost pits and by-products from dairy unit like dung can be used as fish feed or raw material for vermicompost unit.

Therefore, management decisions related to one component may affect the others. Combining ecological sustainability and economic viability, the integrated livestock farming system maintains and improves agricultural productivity while also reducing negative environmental impacts.

Integrated farming system represents an appropriate combination of farm enterprises (cropping systems, horticulture, livestock, fishery, forestry, apiculture, etc.) and the means available to the farmer to raise them for profitability. It interacts adequately with environment without dislocating the ecological and socioeconomic balance on one hand and attempts to meet the national goals on the other (Jayanthi et al., 2002). Thus, integrated farming is a system which tries to imitate the nature's principle, where not only crops but, varied types of plants, animals, birds, fish and other aquatic flora and fauna are utilised for production. These are combined in such a way and proportion that each element helps the other; the waste of one is recycled as resource for the other through 'low external input and sustainable agriculture' techniques (Das, 2013).

Okigbo (1995) defines IFS as a mixed farming system that consists of at least two separate but logically interdependent parts of a crop and livestock enterprises. Enterprises in the integrated farming system are mutually supportive and depend on each other. Mixed farming system consists of components such as crops and livestock that coexist independently from each other. In this farming, integrating crops and livestock serves primarily to minimise the risk and not to recycle resources. But in IFS, crops and livestock interact to create a synergy, with recycling allowing the maximum use of available resources (Sasikala et al., 2015). Crop residues can be used for animal feed, while livestock and livestock by-product production and processing can enhance agricultural productivity by intensifying nutrients that improve soil fertility, reducing the use of chemical fertilisers. A high integration of crops and livestock is often considered as a step forward, but small farmers need to have sufficient access to knowledge, assets and inputs to manage this system in a way that is economically and environmentally sustainable over the long term (FAO, 2001). Thus, IFS is an integrated set of elements/components and activities that farmers perform in their farms under their resources and circumstances to maximise the productivity and net farm income on a sustainable basis (Singh and Ratan, 2009). The integration is made in such a way that the product, i.e. output of one enterprise/ component should be the input for the other enterprises with high degree of complementary effects. The rationale of IFS is to minimise the wastes from the various subsystems on the farm and thus it improves employment opportunities, nutritional security and income of the rural people (Panke et al., 2010).

8.2 SCOPE OF IFS

A combination of one or more farming enterprises with cropping, when carefully chosen, planned and executed, gives greater dividends than a single enterprise, especially for small and marginal farmers. Farm as a unit is to be considered and planned for effective integration of the enterprises to be combined with crop production activity (CARDI, 2010). Integration of farm enterprises depends on the following factors.

1. Soil and climatic features of the selected area
2. Availability of resources, land, labour and capital
3. Present level of utilisation of resources
4. Economics of proposed integrated farming system
5. Managerial skill of the farmer

There are a number of situations and conditions that can be alleviated by an integrated farming system. The following situations are ideal for the introduction of IFS (CARDI, 2010).

1. The farmer wishes to improve the soil quality
2. The farm household is struggling to buy food or below the poverty line
3. Water is stored on-farm in ponds or river-charged overflow areas
4. Fertilisers are expensive or the recommended blend is unavailable
5. Soil salinity has increased as a result of inorganic fertiliser use
6. The farmer is seeking to maximise profits on existing holding
7. The farm is being eroded by wind or water
8. The farmer is looking to reduce chemical control methods
9. The farmer wants to reduce pollution or waste disposal costs

Judicious combination of enterprises, keeping in view of the environmental conditions of a locality will pay greater dividends. At the same time, it will also promote effective recycling of residues/wastes. A grower can use IFS to

1. improve productivity,
2. regulate nutrient and material flows,
3. increase on-farm biodiversity,
4. limit disease, and
5. reduce the smell of some livestock operations.

8.3 GOALS OF FARMING SYSTEMS RESEARCH

Farming system research aims to meet the following interrelated goals (Plucknett et al., 1987).

1. To understand the physical and socioeconomic environment within which agricultural production takes place.
2. To gain an understanding of the farmer in terms of his or her skills, constraints, preferences and aspirations.
3. To comprehend and evaluate existing important farming system, in particular the practice and performance of these systems.
4. To improve the identification of problem and opportunities for change in existing farming system and thereby better focus research on specific key aspects.
5. To enhance the capacity of research organisations to conduct research on priority problems.
6. To conduct research on new or improved practices or principles and to evaluate these for possible testing on-farms.
7. To evaluate new or improved systems or system components on-farms in major production areas under normal farm conditions.
8. To assist the extension, monitor the adoption and assess the benefits of improved farming systems.

8.4 OBJECTIVES OF IFS

The overall objective of integrated farming systems is to evolve technically feasible and economically viable farming system models by integrating cropping with allied enterprises for irrigated, rainfed, hilly and coastal areas with a view to generate income and employment from the farm. The major objectives of integrated farming systems can be listed as below (Behera, 2013).

1. Maximisation of yield of all component enterprises to provide steady and stable income at higher levels.
2. Rejuvenation/amelioration of system's productivity and achieve agroecological equilibrium.
3. Control the build-up of insect pests, diseases and weed population through natural cropping system management and keep them at low level of intensity.
4. Reduction in the use of chemical fertilisers and other harmful agrochemicals and pesticides to provide pollution free, healthy produce and environment to the society at large.

5. Utilisation and conservation of available resources and effective recycling of farm residues within system and to maintain sustainable production system without damaging resources/environment.

8.5 ADVANTAGES OF IFS

Farming systems approach introduces a change in farming techniques for higher production from the farm as a whole with the integration of all the enterprises. The farm products other than the economic products, for which crops are grown, can be better utilised for productive purposes in the farming systems approach. A judicious mix of cropping system with associated enterprises like dairy, poultry, fishery, apiary, sericulture, piggery, etc. suited to the given agroclimatic conditions and socioeconomic status of farmers would bring prosperity to farmer.

Integrated farming system has the advantages of increasing economic yield per unit area per unit time, profitability, sustainability and provides balanced nutritious food for the farmers, pollution free environment and provide opportunity for effective recycling of one product as input to other component, money round the year and solve the energy, fodder, fuel and timber crisis, avoids degradation of forests and enhance the employment generation, increase input use efficiency and finally improve the livelihood of the farming community. Thus the IFS is part of the strategy to ensure sustainable use of the natural resources for the benefit of present and future generations. The major benefits of integrated farming system can be summarised as below.

1. **Productivity:** IFS improves space utilisation and increase productivity per unit area. Thus it provides an opportunity to increase economic yield per unit area per unit time by virtue of intensification of crop and allied enterprises.
2. **Profitability:** In IFS waste materials of one component are used as inputs of another component at the least cost. Thus cost of production is reduced and interference of middleman in procuring most of the inputs is eliminated. Therefore, return per rupee invested or benefit to cost ratio in the farming is increased.
3. **Sustainability:** In integrated farming systems, organic supplementation through effective utilisation of by-products of linked components as a measure is possible and this provides opportunity to promote soil health and to sustain the potentiality of the soil for much longer periods.
4. **Balanced food:** In IFS, components of different nature are linked enabling production of different sources of nutrition, namely, protein, carbohydrates, fats, minerals, vitamins, etc. from the same unit. It provides opportunity to mitigate malnutrition problem of the farm family.

5. **Pollution abatement:** In crop based activity, some of the organics are left as waste materials which in turn pollute the environment on decomposition. Application of huge quantities of fertilisers, pesticides, herbicides, insecticides, etc. pollutes soil, water and air. Much of the wastes could be converted/ recycled to some other forms of economic/ecological/social value, under the integrated farming system. Integrated farming will also help in environmental protection through effective recycling of waste from animal activities like piggery, poultry and pigeon rearing.
6. **Recycling:** Effective recycling of waste materials (crop residues and livestock wastes) is possible in IFS. Therefore, there is less reliance to outside inputs like fertilisers, agrochemicals, feeds, energy, etc.
7. **Income round the year:** IFS provides diversified products. It provides flow of money to the farmer round the year due to integration of enterprises like crops, birds, animals, mushroom, honey, fish, etc. Thus it provides opportunities as crop insurance cover as money round the year. There is higher net return to land and labour resources of the farming family.
8. **Adoption of new technology:** Resources farmer (big farmer) fully utilise technology. On the other hand money flow in IFS round the year gives an inducement to the small/ marginal farmers to go for the adoption of technologies. IFS promotes technology infusion through research and development (R&D) integrated with indigenous/traditional knowledge.
9. **Saving energy:** IFS provides an opportunity to identify an alternative source to the fossil energy. Effective recycling technique of the organic wastes available in the system can be utilised to generate biogas.
10. **Meeting fodder crisis:** Every piece of land area is effectively utilised. Plantation of perennial legume fodder trees on field borders and farm boundary will greatly relieve the problem of non-availability of quality fodder to the farm animals. This practice also improves soil fertility through atmospheric nitrogen fixation. Cultivation of fodder crops as intercropping and as border cropping will also result in the availability of adequate nutritious fodder for animal components like milch cow, goat/sheep, pig and rabbit.
11. **Solving fuel and timber crisis:** The production level of fuel and industrial wood can be enhanced by integrating suitable perennial tree species in farming system (agroforestry) without affecting the crop productivity. This will also greatly reduce pressure on forest resource, thus it preserves the natural ecosystem.
12. **Employment generation:** Integrating crop with livestock enterprises would increase the labour requirement significantly and would help in reducing the

problems of unemployment to a great extent. IFS provides enough scope to employ family labour round the year.

13. **Agroindustries:** Integrated farming also provides opportunities for agri-oriented industries, tourism and related tourism based activities. When produce of any component of the IFS is increased to commercial level there will be surplus for value addition, leading to development of allied agroindustries.
14. **Increasing input efficiency:** IFS provides good scope to use inputs in different components with greater efficiency and benefit cost ratio.
15. **Improving/maintaining soil health:** IFS improves soil fertility and soil physical structure through practice of appropriate crop rotation, cover crop, green manuring, composting farm organic wastes and recycling crop residues. Soil erosion can also be checked by agroforestry and proper cultivation of each part of land by integrated farming.
16. **Reduction of pests:** Adoption of appropriate crop rotation in the farming system can reduce the infestation of weeds, insect pests and diseases.

8.6 COMPONENTS OF IFS

Farmers make decisions about what to grow, what animals to keep, the level and type of inputs and the methods they will use. Their decisions are based upon a range of social, economic and environmental factors. The farmers' attitudes and level of knowledge are also important. However, the selection of enterprises must be based on cardinal principle of minimising the competition and maximising the complementarities between the enterprises (Annadurai et al., 1994). The productivity of farming system depends upon natural resource climate, available science and technology, policies, institutions and public goods, besides latest information of market technology and human resources. The following factors influence the choice of modules in an IFS.

1. Climatic conditions of the locality
2. Soil type of the land
3. Farmers' preferences and household demand
4. Size of the farm
5. Farmer's managerial skill, knowledge and skill
6. Technology for the proposed enterprise available
7. Storage and transport facilities
8. Economical condition of the farmer
9. Resource availability at farm and present level of utilisation of resources
10. Economics of proposed IFS

11. Credit facilities available
12. Social status of the farm family
13. Market facilities available
14. Social customs prevailing in the locality
15. Sentiments and believes of the farm family

The integrated farming system is in vogue from time immemorial, but needs systematic approach starting from diagnosis of the constraints, selection of enterprises to impact analysis in farmers' participatory mode. The focus of present-day agricultural development programmes should be reoriented and changed from commodity based to system-based approach which can gradually be facilitated by capacity building through vocational training and establishment of demonstration units. IFS is an integrated resource management strategy for obtaining economic and sustained crop and livestock production and preserving the resource base with high environmental quality. The entire philosophy of farming system revolves around better utilisation of time, money, resources and family labour. Integrated farming system is focused around a few selected, inter-dependent, interrelated and often interlinking production systems based on few crops, animal and related subsidiary enterprises. It involves the utilisation of primary and/or secondary produce of one system as raw material for other system, thus making the entire system mutually interdependent as one whole unit. In this context there is a necessity to develop models on integrated farming system for different categories of farmers under various agroecological zones. In the process of development of models, the following points should be taken into consideration.

- Sustainability of agriculture depends on the adoption of suitable farming system with inclusion of crop, animal and other allied enterprises.
- Appropriate utilisation of family labour, scientific management of available resources and recycling of agricultural waste by way of integrating different enterprises will make farming a profitable business.
- The farmers should be empowered with technological knowledge and marketing intelligence.

The integrated farming systems include the following components/modules.

1. Crops and cropping systems
2. Horticulture (vegetables, fruits, floriculture, kitchen/nutritional garden)
3. Livestock (cow, buffalo, sheep, goat, pig, chicken, duck, quail, guinea fowl, etc.)

4. Fishery
5. Agroforestry
6. Boundary plantation
7. Apiary
8. Mushroom cultivation
9. Biogas
10. Composting and vermicomposting
11. Vocational activities like leaf-plate making, bamboo/cane work, chalk/candle making, tailoring, embroidery, food processing, doll making, carpentry, hand paper, packaging, paper packet, spice making, *papad* making, marketing agricultural produce, etc.

The crop activities in the IFS consist of grain crops (rice, wheat, maize, sorghum, pulses, soybeans), oilseeds (groundnut, sesame, rapeseed and mustard, linseed, niger, sunflower, castor), vegetable crops, plantation crops (coconut, arecanut), short duration fruit crops (papaya, banana, citrus, pineapple), root crops (cassava, sweet potato), sugarcane, tree crops (moringa, mulberry, teak, acacia, sissoo) and fodder crops. The selection of crops is dependent on preferences based on family consumption, market, soil type, and rainfall and type of animals raised.

Integrated farming system comprising crop and livestock has been sustainable over the centuries. In this system animals are raised on agricultural wastes and animal power is used for agricultural operations and animal excreta are used as manure and fuel. In coastal areas, the rice + fish system can stabilise the productivity and profitability in rice production areas. Cereal based crops in combination with livestock (poultry, dairy, sheep and goat) systems can improve the profitability of the farmers along with the sustenance of natural resources. The livestock activities in IFS consist of cows, goat, sheep, poultry, ducks, pigs and small ruminants. The selection of livestock is also dependent on preference based on family consumption, potential market and availability of resources. Livestock and livelihoods are intimately related and the ownership of livestock is more egalitarian than that of land (Swaminathan, 2010). The livestock component acts as a stabilising factor in the system, thus needs strengthening of the crop and livestock linkage to enhance the economic viability and sustainability of the farming systems (Baishya et al., 2007).

Crop-livestock integration involves the natural resources (crops, animals, land and water) in which these subsystems and their synergistic interactions have a significant positive and the greater total effect than the sum of their individual efforts. These systems increase crop yields and soil fertility and improve the livelihoods with promotion of stable households and increased economic output. The livestock farming systems can be grouped into small holder livestock production with little and low land and commercial livestock production dominating with poultry and dairy. The livestock production in IFS should conserve the natural resource base and raise the productivity through efficient utilisation, optimised allocation and rational management of available of resources.

Integrated farming systems are often less risky and, if managed efficiently, benefits from synergisms among enterprises, diversity in produce, and environmental soundness (Mahapatra, 2010). On this basis, IFS models have been suggested by several workers for the development of small and marginal farms across the country. Dominant farming systems of the different agroclimatic regions are presented in the following table.

Table 8.1. Dominant farming systems in different agroclimatic regions of the country

Agroclimatic region	States	Most dominate farming systems in the region
Western Himalayas	J & K, Himachal Pradesh, Uttarakhand	1. Crops+dairy 2. Agrihorti system
Eastern Himalayas	Assam, Meghalaya	1. Crops+fish+cattle+piggery 2. Monocropping of rice and maize +piggery+backyard poultry+2-3 cows
Trans Gangetic Plain	Punjab, Haryana	1. Crops+dairy
Upper Gangetic Plain	Uttar Pradesh	1. Crops+dairy
Middle Gangetic Plain	Uttar Pradesh, Bihar	1. Crops+dairy 2. Crops+fishery
Lower Gangetic Plain	West Bengal	1. Crops+dairy 2. Crops+dairy+poultry+ duckery
Easter Plateau & Hills	Chhattisgarh, Jharkhand	1. Crops+dairy 2. Crops+backyard poultry+fishery
Central Plateau & Hills	Madhya Pradesh	1. Crops+dairy
Western Plateau & Hills	Maharashtra	1. Crops+dairy 2. Crops+goatery 3. Crops+horticulture
Southern Plateau & Hills	Andhra Pradesh, Telengana, Tamilnadu, Karnataka	1. Crops+dairy 2. Crops+horticulture 3. Crops+horticulture+dairy
East Coast Plain & Hills	Odisha	1. Crops+dairy 2. Crops+dairy+fishery
West Coast Plain & Hills	Maharashtra, Goa, Kerala	1. Crops+dairy 2. Coconut based homestead farming 3. Rice based farming systems

Western Dry Region	Rajasthan	1. Crops+dairy
Gujarat Plain & Hills	Gujarat	1. Crops+dairy
Islands	Andaman & Nicobar	1. Plantation crops+piggery 2. Crops+cattle+fishery

Source: Singh, 2010

8.7 LOW-COST INTERVENTIONS FOR IFS MODULES

A number of 'no cost' or 'low cost' interventions may be made in the modules of IFS for increasing overall productivity and profitability of the marginal and small farmers.

- Development nutritional garden and growing vegetables throughout the year with HYV and hybrids
- Growing climbing vegetable crops on trellis
- Climbers like bottle gourd and ash gourd trailing to the thatches (roof tops) of residential house and cattle shed
- Pruning of fruit trees like mango, guava and cashew trees, pruned materials used as fuel for cooking, as a substitution to the dry cow dung
- Country bean trailing to the fruit crops during *kharif*
- Use of interspace of fruit plant rows in orchards for fodder or pineapple cultivation
- Yam cultivation in fruit orchards
- *Azolla* culture in small pits for feeding dairy animals and poultry birds
- Feeding animals with the leaves of maize plants after the harvest of green cobs
- After harvest of cabbage and cauliflower the leaves are fed to animals
- Use of pulse by-product as cattle feed
- Artificial insemination of cows
- Introduction of coloured dual purpose poultry breed like Chhabro
- Paddy chaffs used as poultry house bedding material
- Use of chopped inflorescence and pseudo-stem of banana plants as poultry feed
- Use of leaves of winter vegetables like cauliflower, cabbage, radish, etc. as poultry feed
- Use of cow dung as fish feed
- Use of weed biomass for compost and vermicompost preparation

- Composting and vermicomposting of all types of organic farm wastes and by-products
- Value addition like preparation and marketing of ghee, *badi*, *papad*, pickle, chips, etc. out of farm produce

8.8 FRUIT CROPS IN IFS

Fruit cultivation and tree based farming plays very important role in imparting sustainability to the integrated farming system. The advantages of fruit crops in integrated farming system may be summarised as follows.

1. Trees trap moisture from deeper layers of soil during drought due to their deep root system. They tolerate adverse weather conditions better than field crops. If annual crops fail due to adverse weather, tree component imparts stability to farmer's economy.
2. Root system of fruit trees penetrate into very deep in soil layers and thus draw plant nutrients from the zones that is not usually available to the herbaceous crops and would have been lost through leaching.
3. This provides higher system productivity and net return.
4. Trees fulfil diverse needs, viz. food, fuel, fodder, fruit, fibre and timber.
5. Fruit trees utilise the off-season rainfall.
6. Canopy of fruit trees slashes the raindrops and thus soil erosion is reduced in high rainfall areas.
7. Perennial fruit crops in the IFS provide employment opportunity round the year.
8. Pruning of perennial fruit trees provide firewood in rural areas.
9. Fruit trees in the IFS gives aesthetic value.
10. Biodiversity is maintained when perennial fruit crops are integrated in the system.
11. Perennial trees provide shade to household and the microclimate is moderated.

8.8.1 Fruit Crops Suitable for IFS

Perennial fruit crops can be used as hedges, individual trees or live fences in IFS, depending on the needs of the farmer, the nature of the fruit species, and the other components of the system. A number of fruit crops, viz. mango, jackfruit, custard apple, guava, sapota, litchi, ber, tamarind, lime, pomegranate, anola, etc. can be grown under integrated farming systems. Most of these fruit crops take

5-8 years for full canopy coverage. In all plantations, field crops and fodder crops can be grown as intercrop during initial years. Up to 3-4 years there is no restriction for choice of field crops and crops can be chosen depending upon climatic and edaphic conditions. Once the productivity of the intercrops is decreased these should be replaced with shade tolerant crops like arrowroot, turmeric, ginger, mango-ginger (*Curcuma amada*), *Aloe vera*, etc. These shade tolerant crops can be grown up to 8-9 years of fruit tree plantation and thereafter pineapple can be taken as intercrop in plantation.

Short duration fruit crops like papaya, banana, pineapple and perennial fruit crops like pomegranate, lime, custard apple, guava, ber, etc. can be planted on pond dykes, if fishery is a component in the IFS. Similarly, these fruit crops can also be planted on household boundaries.

8.8.2 Management of Fruit Trees in IFS

Certain criteria must be followed for the selection of fruit trees for tree based farming systems.

1. The crop should have a commercial potential at the local and regional level.
2. Its growth habit must be compatible with other crops or trees in the system, for example a crown that allows light to reach the associated crops.
3. Fruits should have the ability for prolonged storage and post-harvest processing and transport.
4. The individual fruit tree should have the potential for high production of fruits or biomass.
5. Fruit trees should not act as a host to pests and diseases that are a threat to the associated crops in the system.
6. These should have deep rather than surface root system.
7. Fruits should have the desirable characteristics for industrial uses, for example, pectin content for jelly making.
8. Fruit crops should possess the easier vegetative propagation methods.
9. Fruits should have good flavour and high quality.
10. There should have easy fruit harvesting methods.
11. Fruit crops should have short period of establishment.
12. There should be low incidence of pests and diseases.
13. Fruit trees should have multiple uses.

Table 8.2. Fruit crops and their popular variety suitable for IFS

Crop	Variety	Spacing(m × m)	Bearing age (years)	Economic life (years)	Yield kg/plant
Mango	Amrapalli, Mallika, Kesar, Neelum, Sindhu, Bombay Green, Totapuri	10 × 10 (tall) 5 m × 5 (dwarf)	2-3	25-30	10 kg in 4th yr to 70 kg at full bearing
Jackfruit	Khajara, Gulabi, Champa, Hazar, NJT 1, NJC 1, Singapore	12 × 12	5-6	15-20	4-8 fruits in the beginning, 60-70 fruit at full bearing
Bael	Narendra Bael 1, Narendra bael 2	10 × 10	3-4	25-30	10 kg in 4th yr 70 kg or more at full bearing
Custard apple	Bangalore Selection, Saharanpur local, Mehboobnagar local	5 × 5	2-3	12-15	2- 3 kg at 3 years and 30 kg at full bearing
Guava	Allahabad Safed, Lucknow 49, Banaras Round	5-8 × 5-8	2	12-15	10 kg in 4th yr to 70 kg at full bearing
Aonla	Chakaiya, Banarasi, Krishna, Kanchan, NA 6, NA 7	7-10 × 7-10	3-4	15-20	5 kg in 4th yr 50 kg or more at full bearing
Pineapple	Queen, Kew, Giant kew, Mauritius	0.7 × 0.6	10-12 months	03-04	800-1200 g fruit/plant
Papaya	Pusa Delicious, Pusa Majesty, Pusa Dwarf, Pusa Nanha, Coorg Honey Dew, Taiwan	1.2-1.8 × 1.2-1.8	06-08 months	02-03	15-25 kg / plant up to 3-4 years
Banana	Dwarf Cavendish, Robusta, Grand Naine	1.8 × 1.8 (dwarf) 2.0 2.5-2.5 × 2.0-2.5 (tall)	12-14 months	02-03	15 -20 kg bunch weight/plant

Optimising the use of light, water and nutrients during the formation and development of the fruit increases the quantity and quality of the fruit produced. Therefore, treatments to improve the capture of solar energy and application of water and plant nutrients will maximise the number of fruits that reach maturity. Abortion of fruits and flowers occurs when the inflorescence does not receive sufficient resources.

The main objective of fruit tree management in an integrated farming system is to maximise the value of fruit production without reducing the agricultural production significantly. The trees are to be managed in such a way that the agricultural crops do not suffer from inadequate levels of solar radiation, plant nutrients and soil moisture due to competition from the fruit crop component. Agricultural production is expected throughout the life of the fruit trees in IFS.

Pruning is the removal of plant parts to improve the form and growth of fruit trees. Branches are removed with minimum damage to cambium or growing tissue so that the wound will close in the shortest period of time and with the least possibility of wound infection. The major objectives of pruning are for training the plant, maintaining plant health, improving the quality of flowers, fruit, foliage and stems, and controlling growth. Pruning also promotes regular flowering and fruiting, and maintains a low crown that facilitates fruit harvest and the control of disease. Old, sick, damaged and diseased branches and sprouts should be pruned regularly. The season for pruning depends on the seasons for tree growth and development and on the climatic conditions of the site. It is better not to prune during the dry season when the crops need protection from the summer sun.

8.9 LIVESTOCK-FISH SYSTEMS

Fish-livestock farming systems are recognised as highly assured technology where predetermined quantum of livestock waste obtained by rearing the livestock in the pond area is applied in pond to raise the fish crop without any other additional supply of nutrients. Both production and processing of livestock generate by-products that can be used for aquaculture. Direct use of livestock production wastes is the most widespread and conventionally recognised type of integrated farming. Production wastes include manure, urine and spilled feed; and they may be used as fresh inputs or be processed in some way before use. Based on the type of livestock used for integration there are many combinations in livestock-fish systems.

8.9.1 Cow-Fish System

Manuring of fishpond by using cow dung is one of the common practices all over the world. A healthy cow excretes over 4000-5000 kg dung, 3500-4000 litre urine on an annual basis. Manuring with cow dung, which is rich in nutrients results in increase of natural food organism and bacteria in fishpond. A unit of 5-6 cows can provide adequate manure for 1 ha of pond. Cowshed should be built close to fishpond to simplify handling of cow manure. A cow requires about 7000-8000 kg of green fodder annually. Grass carp utilises the left over grasses, which are about 2500 kg. Fish also utilises the fine feed which consists of grains wasted by cows. In place of raw cow dung, biogas slurry could be used with equally good production. Twenty to thirty thousand kg of biogas slurry are recycled in 1 ha water area to get over 4000 kg of fish without feed or any fertiliser application. In addition to 9000 litres milk, about 3000-4000 kg fish/ha/year can be harvested from dairy-fish integration.

8.9.2 Pig-Fish System

The waste produced by 30-40 pigs is equivalent to one tonne of ammonium sulphate. Exotic breeds like White Yorkshire, Landrace and Hampshire are reared in pigsty near the fishpond. Depending on the size of the fishponds and their manure requirements, such a system can either be built on the bund dividing two fishponds or on the dry-side of the bund. Pigsties, however, may also be constructed in a nearby place where the urine and dung of pigs are first allowed to the digestion chambers of biogas plants for the production of methane for household use. The liquid manure is then discharged into the fishponds through small ditches running through pond bunds. Alternately, the pig manure may be heaped in localised places of fishponds or may be applied in fishponds by dissolving in water. Pig dung contains more than 70% digestible feed for fish. The undigested solids present in the pig dung also serve as direct food source to tilapia and common carp. Forty pigs are enough to fertilise a fishpond of one hectare area. Fish like grass carp, silver carp and common carp are suitable for integration with pigs.

8.9.3 Chicken-Fish System

Chicken raising for meat or eggs can be integrated with fish culture to reduce costs on fertilisers and feeds for fish and maximise benefits. Chicken (cock and hen) can be raised over or adjacent to the ponds and their excreta recycled to fertilise the fishponds. Chicken housing, when constructed above the water level using bamboo poles would fertilise fishponds directly. In fish-chicken integration, birds housed under intensive system are considered best. About 6-8 cm thick layer prepared from chopped straw, dry leaves, saw dust or groundnut shell is sufficient. Birds' droppings in the form of litter contain 3% nitrogen, 2% phosphate and 2% potash. Thus, it produces phytoplankton and zooplankton in fishpond. So application of extra fertiliser for fish is not required. This cuts the cost of fish production by 60%. In one year 25-30 birds can produce one tonne litter and based on that it is found that 500-600 birds are enough to fertilise one ha water spread area for good fish production. When phytoplanktonic bloom is seen over the surface water of pond then application of birds' litter to the pond should immediately be suspended. Chicken-fish integration also maximises the use of space and saves labour in transporting manure to the ponds.

8.9.4 Duck-Fish System

A fishpond being a semi-closed biological system with several aquatic animals and plants provides excellent disease-free environment for ducks. In return ducks consume juvenile frogs, tadpoles and dragonfly, aquatic weeds, thus making a safe environment for fish. Duck dropping provides essential nutrients to stimulate growth of natural food for fish. This integration is highly profitable as it greatly enhances the animal protein production in terms of fish and duck per unit area. The duck dropping contains 25% organic and 20% inorganic substances with a number of elements such as carbon, phosphorus, potassium, nitrogen, calcium, etc. Hence, it forms a very good source of fish feed. Besides manuring, ducks eradicate the unwanted insects, snails and their larvae which may be the vectors of fish pathogenic organisms and water-borne disease-causing organisms infecting human beings. The number of ducks may be between 200 and 400/ha depending on the duration of fish culture and the manure requirements. For culturing fish with ducks, it is advisable to release fish fingerlings of more than 10 cm size; otherwise the ducks may feed on the fingerlings. As the nitrogen-rich duck manure enhances both phytoplankton and zooplankton production, phytoplankton-feeding silver carp and zooplankton feeding catla and common carp are ideal for duck-fish culture.

8.10 LIVESTOCK-CROP SYSTEMS

An integrated crop-livestock system is a form of mixed production that utilises crops and livestock in a way that they can complement one another through space and time. In livestock-crop system, the animal component is often raised on agricultural waste products while the animal is used to cultivate the land and provide manure to be used as fertiliser and fuel (Jayanthi et al., 2000). The animals like sheep, goats or cattle play important role in enriching the soil gradually and increasing soil organic matter to support the crop. Animal can also be used for farm operations and transport, while crop residues provide fodder for livestock and grain provides supplementary feed for productive animals. Animals also provide manure and other types of animal waste. Cow dung helps in the overall sustainability of the farming system. Dung contains macronutrients and micronutrients required for improving the soil fertility and crop growth. Cow dung is used for the production of biogas. Biogas is a source of renewable, alternative and sustainable energy (Godi et al., 2013).

8.10.1 Small Ruminant-Crop System

Small ruminants play a vital role in the economy of small, marginal and landless farmers. Integration of these enterprises can increase the total productivity, maintain ecological balance and economic sustainability. Small ruminants based integrated farming system also helps in improving the soil fertility. Small ruminants graze weeds in the crop field. They also directly graze fodder tree/shrubs. Thus, this system saves labours and spread of insect pests is checked. Boundary plantation with fodder shrubs/trees provides high quality leaf fodder for small ruminants.

8.10.2 Rice-Duck System

Rice-duck farming is an integrated type of farming technology. It is especially suitable for resource poor farmers to produce organic rice in low cost. The evidence from various countries including Japan, Bangladesh, Philippines, and Vietnam has proved the integration of ducks in rice field as a successful and productive farming technology. The integration of ducks in rice field creates symbiotic relationship between rice and ducks yielding mutual benefits to both entities. Ducks eat harmful insects and weeds averting the use of chemical pesticides and manual weeding in the rice field. Thus they get nutritious diet from eating insects and weeds in rice field. The droplets of ducks act as natural fertiliser to the rice crop preventing the use of chemical fertilisers. The continuous movement of ducks in the rice field provides natural stimulation and aeration which increases the availability of nutrients like nitrogen, phosphorous and potash to the rice crop. Rice-duck technology causes the reduction of emission of methane gas from rice field contributing to reduce the global warming. Against the traditional rice farming system, integrated rice-duck technology supersedes in terms of minimising the cost of production, increasing rice productivity, providing environmental benefits and increasing the income of farmers through sale of organic rice and duck meat.

8.10.3 Livestock-Crop-Fish System

Livestock-crop-fish farming systems can be followed by integrating fish with livestock-crop farming system, without any additional feed to fish, rearing fish in the pond with the help of available dung from livestock. Integrated livestock-crop-fish farming can be carried out for increasing returns from a limited land area and reducing risk by diversifying crops. There should be ample supply of fresh and clean water available throughout the year to maintain water level for fish management purposes. The nutrients content in cow dung will help in growth of phytoplankton and zooplankton in the pond. The by-products of livestock can be used for aquaculture. Direct application of wastes of livestock is common practice. The by-products of livestock are manure and urine.

8.11 FISH-CROP SYSTEMS

8.11.1 Fruit Crops/Vegetables on Pond Dykes

Pond acts as a mini water harvesting structure which holds a considerable amount of rain and runoff water after cessation of rainfall. Farmers can grow crops in the dyke with proper planning and adoption of suitable cropping sequence by utilising the stored water. Fruit crops like coconut, papaya, banana, pomegranate, lemon, and seasonal vegetables like ridge gourd, bottle gourd, bitter gourd, pumpkin, cucumber, etc. can be grown successfully on pond dykes.

8.11.2 Hanging Platforms

Ten to twelve no of bamboo platforms of 4 m × 3.5 m size are constructed from the inner side of the dyke hanging over the trenches for planting the creeper vegetables at the inner side of the dyke and allowing the plants to creep over it. Climbing vegetables such as pumpkin, ridge gourd, and sponge gourd are grown during pre-kharif and kharif season. During rabi season ash gourd, bottle gourd and country bean are grown. These climbing vegetable platforms provide shade to the fishes during hot day in addition to production of 40-50 kg of vegetables/ season/platform.

8.12 MUSHROOMS IN IFS

There are many reasons why mushroom agriculture is more sustainable than other types of agriculture. Farmers use by-products from other agricultural production systems such as cow manure, chicken litter, cotton hulls, paddy straw, etc. for substrate on which mushrooms grow. Mushroom cultivation activities can play an important role in supporting the local economy by contributing to subsistence food security, nutrition, and medicine; generating additional employment and income through local, regional, and national trade; and offering opportunities for processing enterprises (Marshall and Nair, 2009). Mushroom cultivation contributes to strengthening the livelihoods through reducing vulnerability to poverty and generating reliable sources of income. The demands of mushroom in pharmaceutical industry increase due to its nutraceutical application. Mushrooms are being used for the production of various value-added products such as mushroom *papad*, mushroom pickle, mushroom biscuits, mushroom beverage, etc.

8.12.1 Mushroom Compost in Horticulture

Spent mushroom compost (SMC) is the residual compost waste generated or left out after mushroom productions. This compost is readily available and its formulation generally consists of a combination of cereal straw, animal manure, ground chalk, etc. composted all together. SMC contains lots of salt and organic material along with enzymes and other nutrients that make it suitable habitats for different microbes including bacteria and fungi, synergistically they are found to be beneficial in disease suppression and plant growth promotion. Disease suppressive properties of composts rely on several factors including microbial activity, microbial population dynamics, nutrient concentrations, and other associated chemical and physical factors (Patil et al., 2018). Use of SMC as a soil amendment shows positive and significant effects on crop growth parameters. Spent mushroom substrates (SMS) are those residues which are left after the mushroom harvest from the non-composted substrate. Both SMS and SMC have a common role in the soil as a soil conditioner. These improve soil structure, provide plant nutrients, increase plant nutrient availability, increase soil microbial populations, increase soil cation exchange capacity, increase plant root structure, increase soil aeration, improve soil water status, and reduce soil compaction. Therefore, SMC and SMS are suitable to use in high-value horticultural crops (Singh et al., 2020).

8.12.2 Mushroom in Animal Husbandry

Some species of edible mushrooms as a feed additive to the broiler chickens and laying hens have dietary and health-promoting activities (Bederska-Lojewska et al., 2017). In poultry farm, antibiotic is common to increase the production which leads to human health hazards. Therefore, a natural source of the antibiotic substitute in the form of medicinal mushrooms shows positive effects on immunity on the broiler and laying hens (Mahfuz and Piao, 2019). The waste of oyster mushroom left from the mushroom industry may be used as supplements to the broiler chicken (Fard et al., 2014).

8.13 AGROFORESTRY SYSTEMS

Agroforestry is a sustainable land management system that increases overall production, combines agricultural crops, tree crops and forest plants and/or animals simultaneously or sequentially and applies management practices that are compatible with cultural patterns of local population. In other words, agroforestry is the deliberate growing of woody perennials on the same unit of land along with agricultural crops and/or animals, either in some form of spatial mixture or in some temporal sequence where both ecological and economical interactions exist between the different components. Based on nature of components, agroforestry systems may be agrisilvicultural, silvipastoral, agrisilvipastoral, hortisilvicultural, hortisilvipastoral, agrihortisilvipastoral, aquaforestry systems, etc. One or more of these systems can be integrated to various farming systems.

8.13.1 Agrisilvicultural Systems

In these systems, annual agricultural crops and perennial woody trees/shrubs are integrated. Based on the nature of the components this system can further be grouped into various forms such as alley cropping, shelterbelts, windbreaks, etc.

Alley cropping or hedgerow intercropping is a management-intensive agroforestry practice in which perennial, preferably leguminous trees or shrubs are grown simultaneously with an arable crop. The trees, managed as hedgerows, are grown in wide rows and the crop is planted in the interspaces or 'alley' between the tree rows. The trees are pruned regularly during the cropping phase and allowed to grow freely to shade the inter-rows when there are no crops. Alley cropping retains the basic restorative attributes of the bush fallow through nutrient recycling, fertility regeneration and weeds suppression and combines these with arable cropping so that all processes occur concurrently on the same land, allowing the farmer to crop the land for an extended period.

In this system, trees are usually planted 6-10 m × 3 m spacing and in alleys suitable arable crops are grown. For first 3-4 years there are no restrictions for choice of annual crops as there is not much competition for light among the trees and the herbaceous crops. But as canopy closes the shade tolerant crops like ginger, turmeric and arrowroot may be taken from 4th to 6/7th year of tree plantation and then from 6/7th year medicinal plants like kalmegh, patchouli and *Aleo vera* or fruit crops like pineapple may be taken successfully. Tree crops may be harvested depending upon their use and species.

The choice of tree species for alley cropping is extremely important and to a large extent determines the success or failure of the system. The tree species for an ideal alley cropping should have rapid growth rate, ability to withstand frequent cutting, good coppicing ability, easy to establish from seeds or cuttings, nitrogen fixing capacity, deep-rooted with a different root distribution to the crop, and ability to withstand environmental stresses. The suitable species for alley cropping are *Cassia siamea*, *Leucaena leucocephala*, *Gliricidia sepium*,

Calliandra calothyrsus, *Sesbania sesban*, *Acacia mangium*, *Gmelina arborea*, *Albizia* spp., *Cajanus* spp., *Dalbergia sissoo*, *Tectona grandis*, *Chamaecytisus* spp., *Desmodium* spp., *Erythrina* spp., *Flemingia* spp., *Inga* spp. and *Tephrosia* spp.

The agrihortisilvicultural system is similar to agrisilvicultural system, but here trees are usually planted 6-10 m × 6 m spacing. In between two trees one fruit crop like mango, guava or pomegranate may be planted. Suitable arable crops are grown in inter-spaces of perennial fruit + multipurpose tree rows. When annual crop yield is substantially reduced due to canopy closure the system is converted to a hortisilvicultural system. With appropriate management practices this practice can give a sustainable production.

8.13.2 Silvipastoral Systems

Silvipasture as an agroforestry practice is specifically designed and managed for the production of trees, tree products, forage and livestock. Silvipasture results when forage crops are deliberately introduced or enhanced in a timber production system, or timber crops are deliberately introduced or enhanced in a forage production system. Integrating livestock production in the farming system promotes farm product diversification, improves food security and offers a diversified marketing opportunity that can stimulate rural economic development. The interactions among timber, forage and livestock are managed intensively to simultaneously produce timber commodities, a high-quality forage resource and efficient livestock production. Overall silvipastoral systems can provide economic returns while creating a sustainable system with many environmental benefits.

Before new silvipastoral systems are established, implications of merging forestry and agricultural systems should be explored thoroughly for economic and environmental considerations. Factors like availability of potential markets, soil type, climatic conditions and species compatibility should be considered while choosing tree and forage crop species. The timber component should be marketable, of high quality, fast growing, deep rooted, drought tolerant, and capable of providing the desired products and environmental services. The forage component should be a perennial crop that is suitable for livestock grazing, compatible with the site (soil, temperature, precipitation), productive under partial shade and moisture stress, responsive to intensive management and tolerant of heavy utilisation or grazing. A successful silvipastoral system requires understanding forage growth characteristics and managing the timing and duration of grazing to avoid browsing of young tree seedlings. Improper management of silvipastures can reduce desirable woody and herbaceous plants by overgrazing and soil compaction.

In silvipastoral and hortipastoral systems, suitable fodder grasses like guinea (*Panicum maximum*), dinanath grass (*Pennisetum pedicellatum*), napier grass (*Pennisetum purpureum*), thin napier (*Panicum polystachion*), anjan grass (*Cenchrus ciliaris*) and fodder legumes like stylo (*Stylosanthes hamata* and *Stylosanthes scabra*), cowpea (*Vigna sinensis*) are grown in inter-spaces of perennial fruit or multipurpose tree rows. These systems are very effective to control soil erosion as the grasses give a permanent soil cover.

Within the broad category of silvipastoral system there are several types of practices which can be identified depending on the role of the tree/shrub component. These practices include protein banks (fodder tree banks), trees and shrubs on rangeland or pastures, live fences of fodder trees and shrubs and plantation crops with pastures and animals.

8.13.3 Agrisilvipastoral Systems

Agrisilvipastoral system is the combination of agricultural crops, woody perennials and livestock. Agroforestry practices under this system include agrisilvicultural system converted to silvipastoral systems, multistorey system with free grazing, homegardens and alley cropping system using pasture grasses/fodder crops and agricultural crops.

In agrisilvicultural system converted to silvipastoral systems, the initial cropping combinations include tree seedlings and annual agricultural crops. As the trees grow and close their canopies, the yield recovery of annual agricultural crops gradually decreases and it will no longer be possible to grow these crops. Instead, shade tolerant grasses and vines will be more compatible to land use system where animals are allowed to graze freely, thus converting agrisilvicultural system to silvipastoral system. Multistorey system with free grazing is similar to the multistorey under agrisilvicultural system, except that in this case, grazing animals are an added component. Alley cropping with pasture grasses and agricultural crops is similar to hedgerow cropping with seasonal agricultural crops. However, instead of all alleys planted with agricultural crops, some alleys in between the hedgerows are grown with improved pasture grasses and/or fodder trees or shrubs which are regularly cut and fed to livestock. In woody hedgerows, various woody hedges especially fast growing and coppicing fodder shrubs and trees are planted for the purpose of browse, mulch, green manure, soil conservation, etc. The species for this system are *Erythrina* spp., *Leucaena luecocephala*, *Sesbania grandiflora*, etc.

8.13.4 Homegardens

Homegarden is one of the oldest agroforestry practices found extensively in the high rainfall areas in tropical south and south-east Asia. The word 'homegarden' has been used to describe diverse practices from growing vegetables behind houses to complex multistorey systems. It is the intimate association of multipurpose trees and shrubs with seasonal and permanent arable crops and livestock within the compounds of individual houses. Mostly the whole system, i.e. crop-tree-animal unit is managed by family labour. These systems are common in all ecological regions in the tropics and subtropics, especially in humid lowlands with high population density. The average size of a homegarden is usually much less than one hectare. In India, every homestead has around 0.2 to 0.5 ha land for personal production. In many parts of the world the fruit, vegetable and other foodstuff produced from homegardens provide a substantial part of the household food requirement. Food production is the primary function and role of most of the homegardens. The combination of crops with different production cycles results in a relatively uninterrupted supply of food products throughout the year. In most cases, animals and birds are kept in the homegardens. Fodder and legumes are widely grown to meet the daily fodder requirement of the animals. The waste materials from crops and homes are also used as fodder for animals and feed for birds. Cattle are mostly kept for dairy products and land cultivation. Goats, chickens and fish are kept for household consumption. Products from animals or the animals themselves can also be sold in the market.

The layered configurations and combination of compatible species are the most important characteristics of all homegardens. The gardens are carefully structured systems where each component has a specific place and function. Every homegarden usually consists of an herbaceous layer near the ground, a tree layer at the upper level, and intermediate layers in between. The lower layer is usually divided into two, with the lowermost (less than 1 m height) dominated by different vegetable and medicinal plants, and the second layer (1-3 m height) being composed of food plants such as banana, papaya, yam and so on. The upper tree layer is also partitioned into two, consisting of emergent, fully grown timber and fruit trees occupying the uppermost layer of over 25 m height, and medium-sized trees of 10-20 m height occupying the next lower layer. The intermediate layer of 3-10 m height is dominated by various fruit trees.

8.13.5 Aquaforestry

The aquaforestry system comprises of composite fish culture in farm ponds, and various trees and shrubs (*Leucaena leucocephala*, *Morus alba*, *Gliricidia sepium*, *Moringa olifera*, etc.), leaves of which are preferred by fish are planted on the boundary and around fishponds. Leaves of these trees are used as feed for fish. Inland fish such as catla, rohu, mrigal, common carp, silver carp and grass carp can be grown in the ponds. In the coastal regions, farmers are cultivating fish and prawn in saline water and growing coconut and other trees on bunds of ponds. Now fish culture in the mangroves is also advocated, which form a rich source of nutrition to the aquatic life and breeding ground for fish and prawn. Other terms that have been used for this practice are silvipisciculture, agrisilviaquaculture and aquasilviculture. Some other modified aquaforestry systems are as follows.

1. A well-balanced system of animal husbandry including goatery, poultry, duck farming, turtles and fishes in the small ponds in homegardens make a balanced system of high moisture, energy and nutrient use efficiency per unit area.
2. In paddy field, fish can easily be reared by planting trees on field bunds or boundary to provide leaves used as fish feed. This system can be practised in high rainfall areas.
3. Coconut plants can also be successfully planted on raised paddy field bunds with an alley space of 5 m width depression which is utilised for pisciculture purpose.

8.13.6 Apisilviculture

In this system, various honey or nectar producing trees frequently visited by honeybees are planted on the boundary of the agricultural field. Bees benefit the trees and trees in turn provide series of benefits to bees. The primary purpose of this system is to produce honey. Apisilviculture with *Eucalyptus*, *Gliricidia*, *Grevillea*, *Gmelina*, *Leuceana* and *Albizia* species are more remunerative and a good source of generating additional farm income in rural areas. Fruit trees can also be integrated with apisilviculture for more profits.

8.14 IMPACT OF FARMING SYSTEMS ON ENVIRONMENT

Agriculture is the world's largest industry. It employs more than one billion people. Pasture and cropland occupy around 50% of the earth's habitable land and provide habitat and food for a multitude of species. On the other hand, modern agriculture and food production contribute a significant share of the greenhouse gas (GHG) emissions that are causing climate change; 17% directly through agricultural activities and an additional 7-14% through changes in land use. In agriculture, the non-CO_2 sources (CH_4 and N_2O) are reported as anthropogenic GHG emissions. The CO_2 emitted is considered neutral, being associated to annual cycles of carbon fixation and oxidation through photosynthesis. Pesticides, fertilisers and other toxic farm chemicals can also poison fresh water, marine ecosystems, air and soil. They can remain in the environment for generations.

When agricultural operations are sustainably managed, they can preserve and restore critical habitats, help protect watersheds, and improve soil health and water quality. But unsustainable practices have serious impacts on people and the environment. The need for sustainable resource management is increasingly urgent. Demand for agricultural commodities is rising rapidly as the world's population grows.

8.14.1 Effect of Cropping Systems on Environment

8.14.1.1 Methane emission from rice fields

The rice fields are a major source of emission of GHGs like CH_4 and N_2O. Methane is produced as the terminal step of the anaerobic breakdown of organic matter in wetland rice soils. The oxygen supply from the atmosphere to the soil of a flooded rice field is cut off which leads to anaerobic fermentation of soil organic matter. Methane is a major end-product of anaerobic fermentation. It is released from submerged soils to the atmosphere by diffusion and ebullition and through roots and stems of rice plants. Disturbance of wetland soil by cultural practices favours soil trapped methane to escape to atmosphere through ebullition. Methane is exclusively produced by methanogenic bacteria that can metabolise only in the strict absence of free oxygen. Though flooded paddy soils have a high potential to produce CH_4, but part of produced CH_4 is consumed by CH_4 oxidising bacteria or methanotrophs. In rice fields, it is possible that a part of produced CH_4 in anaerobic soil layer is oxidised in aerobic layers such as surface soil-water interface and the rhizosphere of rice plants, and the net emission will be positive or negative depending on the relative magnitudes of methanogenesis and methanotrophy, respectively (Lenka et al., 2015).

In tropical flooded rice soils, where soil temperature is 25-30° C, methane production is rapid in alkaline and calcareous soils, may start hours after flooding and slow in acid soils, and may take five or more weeks after flooding. Methane production is negatively correlated with a soil redox potential and positively correlated with soil temperature, soil carbon content, and rice growth. Easily degradable crop residues, fallow weeds, and soil organic matter are the major source for initial methane production. At later growth stages of rice, root exudates, decaying roots, and aquatic biomass are more important for CH_4 emission. Use of organic amendments and chemical fertilisers also promotes CH_4 emission.

8.14.1.2 N_2O emission from agricultural soils

Most of the N_2O emissions take place in soils and are related with agricultural activities. The N_2O emitted from the soil represents some 50% of the total agricultural emissions. Even when it is not being cultivated, the soil naturally releases GHGs. The N_2O is generated as a by-product of microbial activities that convert ammonium into nitrate (nitrification) or nitrate into nitrogen gas (denitrification). Both processes are influenced and controlled by environmental conditions such as soil moisture, temperature, oxygen concentration, amount of available organic carbon and nitrogen and soil C/N ratio. Emissions also increase with agricultural activity, partly as a result of N input from manure, mineral fertilisers, or from symbiotic N fixation in legumes. Among these factors, those related to soil could be easily altered by management practices. Therefore, understanding the processes of N_2O formation in soils and the factors influencing these emissions is fundamental to develop efficient strategies to reduce N_2O emissions in agricultural soils (Signor and Cerri, 2013).

8.14.1.3 Burning of agricultural residues in field

In developing countries, farmers prefer crop residue burning as a quick and labour-saving process to dispose of the crop residues of rice, wheat, maize and sugarcane. Emissions of CO_2 during burning of crop residues are considered neutral, as it is reabsorbed during the next growing season. However, biomass burning is one of the significant sources of atmospheric aerosols and trace gas emissions, which has a major impact on human health. In addition to aerosol particles, biomass burning due to forest fires and crop residue burning are considered a major source of carbon dioxide, carbon monoxide (CO), methane, volatile organic compounds (VOC), nitrogen oxides, and halogen compounds. Carbon monoxide is a chemically active gas in the troposphere influencing the abundance of O_3 and the oxidising capacity of the troposphere. Thus, an increase in concentration of CO, VOC, and nitrogen oxides also increases concentration of GHGs in the atmosphere (Lenka et al., 2015).

8.14.1.4 Use of fossil fuels in agriculture

Use of fossil fuels in agriculture results in CO_2 emissions, and there are additional emissions associated with production and delivery of fuels to the farm. Non-traditional fuels sometimes used in processing agricultural materials include scrap tires and biomass. The CO_2 emission attributed to electricity consumption is based on the fuels used in power generation.

8.14.1.5 Use of fertilisers

Fertiliser production consumes about 1.2% of the world's energy and is responsible for approximately 1.2% of the total GHG emissions. Carbon dioxide emissions result from the energy required for production of fertilisers and the energy required for their transport and application. The energy required per tonne of N and phosphate varies considerably with the form in which the nutrient is supplied. Carbon emissions from fossil fuels used in the production of fertilisers include emissions from mineral extraction and fertiliser manufacture. Postproduction emissions can include those from packaging, and transportation. Energy is also used during field application of fertilisers using farm machinery, thus the greater the fertiliser use, the greater are the emissions. Carbon emissions from agricultural lime are calculated from the fuel used for mining limestone and for grinding the stone into a usable product.

8.14.1.6 Pesticides

Modern pesticides are almost entirely produced from crude petroleum or natural gas products. The total energy input is thus both the material used as feedstock and the direct energy inputs. Carbon dioxide emissions from production of pesticides consist of both these contributions to manufacture the active ingredient. Postproduction emissions include those from formulation of the active ingredients into emulsifiable oils, wettable powders, or granules and those from packaging, transportation, and application of the pesticide formulation (Lenka et al., 2015).

8.14.1.7 Irrigation

The on-farm wells, on-farm surface reservoirs, and off-farm surface reservoirs are the major sources of water for farm activities. Fossil fuels used to power pumps, which distribute irrigation water. The energy use and carbon emissions from pumping water are applied to both on-farm wells and off-farm surface reservoirs.

8.14.1.8 Harvesting and threshing

Energy and CO_2 emissions during harvesting and threshing of agricultural produce are also important. The greater the productivity, the greater are the energy and emissions required for harvesting and threshing operations.

8.14.1.9 Farm machinery

Energy and CO_2 emissions associated with different tillage practices are a consequence of the fuel used by farm machines and the energy consumed in manufacture, transportation and repair of these machines.

8.14.1.10 Forestry and other land use changes

The expansion of agricultural land area has a major environmental impact. Deforestation and depletion of the humus releases large quantities of CO_2 from the carbon bound in the trees and the soil organic matter. Furthermore, deforestation has an immediate impact on the natural water cycle, resulting in a greater likelihood of flooding or drought. Some 24% of the total global GHG emissions can be attributed to agriculture and about 12% of these are due to change in land use and with extended agricultural production, this would rise considerably (FAOSTAT, 2013). Changes in land use have negatively affected the net ability of ecosystems to sequester C from the atmosphere. For instance, the C-rich grasslands and forests in temperate zones have been replaced by crops with much lower capacity to sequester C (Lenka et al., 2015).

8.14.2 Effect of Livestock on Environment

Ruminant animals, such as cattle, sheep, buffalo, and goats have special digestive systems that can convert otherwise unusable plant materials into nutritious food and fibre. This helpful digestive process, called enteric fermentation, however, produces methane, a potent greenhouse gas that can contribute to global climate change. Enteric fermentation is a digestive process by which carbohydrates are broken down by microorganisms into simple molecules for absorption into the bloodstream of an animal. It is one of the factors in increased methane emissions. Enteric fermentation occurs when methane is produced in the rumen as microbial fermentation takes place.

Livestock production systems can also emit other greenhouse gases such as nitrous oxide and carbon dioxide. In recent years, the increasing use of intensive livestock production systems has become a source of solid, liquid and airborne emissions that can be both a nuisance and environmentally harmful. In spite of the low amount of CH_4 in the atmosphere relative to that of CO_2, its importance as a pollutant is considered to be 21 times greater than that of CO_2, while that of N_2O is 310 times greater (Hartung, 2003).

It is estimated th0at nearly 20% of CH_4 and 35% of N_2O come from livestock production. According to the European Environment Agency, nearly 50% of the overall amount of CH_4 released in Europe originates from agriculture and stems mainly from ruminant animals.

Manure of cattle is a much more important source of CH_4 emission than enteric fermentation. This problem is due to liquid manure storage in tanks or pits. The amount and surface area of the stored manure, and the ambient and core manure temperature are the major factors that determine the quantity of CH_4 gas emission.

Thus, the livestock sector which will be a sufferer of climate change is itself a large source of methane emissions. In India, although the emission rate per animal is much lower than the developed countries, due to vast livestock population the total annual methane emissions from Indian livestock ranged from 7.26 to 10.4 million tons/year. In India more than 90% of the total methane emission from enteric fermentation is being contributed by the large ruminants and rest from small ruminants and others.

8.14.3 Agricultural Waste and the Environment

Agricultural waste is now considered as an important source of environmental pollution. Huge quantities agricultural wastes from the livestock farms and agroindustries are entering into the environment. The impact of agricultural waste on the environment depends not only on the quantity generated but also on the disposal practices followed. Some of the disposal methods pollute the environment. Cereal straw burning is a common practice in the developing countries which generates a lot of harmful gas, smoke and dust. These pollutants are accompanied by the formation of ozone and nitric acid and contribute to acid deposition which poses risk to human health and ecological environment. Agricultural residues are also directly discharged into water, leading to serious contamination of aquatic environment (Wanga et al., 2016).

Animal wastes such as faeces, urine, and respiration and fermentation gases cause an acute and serious environmental problem in countries with high concentrations of animals on a limited land base for manure disposal. Animal wastes are excreted in solid, liquid, and gaseous forms. Respiration and fermentation gases are lost to the environment soon after being produced by the animal. After excretion, solid and liquid animal waste is subjected to microbial conversion, mainly through anaerobic process, which converts organic substrates into microbial biomass and soluble and gaseous products. Some of these products have an impact on the environment, as well as water quality, soil deterioration, and air pollution (Sabiiti, 2011). Animal manure also contains many pathogens, parasite eggs, heavy metals, etc. Application of excessive animal wastes on land as fertiliser and soil conditioner is subject to surface runoff and leaching that may contaminate ground or surface water. Nitrate leaching is considered a major nitrogen pollution concern on livestock farms (Mackie et al., 1998). When phosphorus enters the surface water from land application of excessive animal manure it can stimulate the growth of algae and other aquatic plants. Their subsequent decomposition results in an increased oxygen demand that interferes with the fish and other aquatic animals. Manure decomposition can be a major source of methane, ammonia and nitrogen oxides, which contribute to accumulation of greenhouse gases. Volatilisation of ammonia causes acid deposition, which contributes to acid precipitation (Likens et al., 1996). Odour pollution is also reported due to improper disposal of livestock wastes.

8.15 RECYCLING AND UTILISATION OF AGRICULTURAL WASTES

All kinds of agricultural wastes, especially poultry, animal faeces and crops straw, have a very high nutrition potential and can also improve soils for sustainable production ability (Ibrahim, 2015). Therefore, the effective transformation of agricultural waste recycling and utilisation was important in the controlling of environment pollution. Besides, facing with the problem, it can also address the serious energy crisis. The recycling and utilisation of agricultural wastes are considered to be the important step in environmental protection, energy structure and agricultural development (Wanga et al., 2016). Usually, agricultural wastes are utilised in the traditional ways with the low utilisation efficiency. The disposal and utilisation of agricultural wastes depend upon the availability of viable technology and agricultural automation processes, government policies, rules and regulations, and social service systems available in agricultural waste management (Nguyen et al., 2014).

The pollutants from agriculture wastes have four characteristics, including huge quantities, bad qualities, low price and excessive danger (Rangabhashiyam et al., 2013). However, the target should be to make agricultural wastes a resource that can be utilised and not just discarded. Agricultural wastes can be used to enhance food security mainly through their use as biofertiliser and soil amendment, use as animal feed, and energy production. They contain large amounts of organic matter, and many of them can be directly added to the soil without any risk (Sabiiti, 2011).

8.15.1 Agricultural Wastes Utilisation Practices

Agricultural waste utilisation technology must either use the residues rapidly, or store the residues under conditions that do not cause spoilage or render the residues unsuitable for processing to the desired end-product. There are a number of applications to which these wastes can be used.

8.15.1.1 Organic fertilisers

The utilisation of animal manures for plant nutrients has a definite impact on input energy requirements at the farm level. Converting crop residues and animal manures into organic fertilisers through composting is one of the waste treatment technologies that make it possible to use organic waste as plant nutrients. Composting reduces the volume of the waste, hence solving serious environmental problems concerning disposal of large quantities of waste, kills pathogens that may be present, decreases the germination of weeds in agricultural fields, and reduces odour. Poultry manure contains high phosphorus which has positive effect on the growth and productivity of crops. It is also effective when combined with mineral phosphorus fertiliser for farm use. Adding organic manure to soil increases its fertility because it increases the nutrient retention capacity, improves the physical condition, the water holding capacity and the soil structure stability.

8.15.1.2 Animal feed

Both crop residues and animal waste can be used as animal feed. Crop residues are high in fibre content but low in protein, starch and fat content. To increase the livestock production the forage and pasture should be supplemented with grains and protein concentrate to meet the protein needs of the animals. Thus, use of the grain and protein for human food will compete with such use for animal feed. These problems may be overcome by utilising crop residues to feed the animals. The nutrient content of animal waste depends on the animal species, type of feed, and bedding material used. The ruminants are useful in converting crop residues into food, hence contributing substantially to reducing potential pollutants. The rumen contains the microbial enzyme cellulase, which is the only enzyme to digest the most abundant plant product, cellulose. With ruminants, nutrients in crop by-products are utilised and do not become a waste-disposal problem.

8.15.1.3 Anaerobic digestion

Methane gas can be produced from agricultural wastes particularly manures. The gas is best suited for water heating, grain drying, etc. The anaerobic digestion of agricultural waste to form methane-rich gas is a two-step microbial fermentation. Initially, acid-forming bacteria break down the volatile solids to organic acids which are then utilised by methanogenic organisms to yield methane-rich gas. The composition of the typical gas produced is methane 50-70%; CO_2 25-45%; N_2 0.5-3%; H_2 1-10% with traces of H_2S; and the heating value of the gas is in the range of 18-25 MJ/m^3. Some of the major disadvantages of the digestion system are the high capital costs and the explosive properties of the methane gas. However, the advantages far outweigh the disadvantages. Anaerobic digestion makes the treatment and disposal of large poultry, swine and dairy waste feasible, minimising the odour problem. It stabilises the waste and the digestion sludge is relatively odour-free and yet retains the fertiliser value of the original waste (Obi et al., 2016).

8.15.1.4 Adsorbents in the elimination of heavy metals

Excessive release of heavy metals into the environment due to industrialisation and urbanisation has posed a great problem worldwide. Unlike organic pollutants, the majority of which are susceptible to biological degradation, heavy metal ions such as copper, cadmium, mercury, zinc, chromium and lead ions do not degrade into harmless end-products. The presence of heavy metal ions is a major concern due to their toxicity to many life forms. It is reported that use of agricultural wastes is a low cost alternative for the treatment of effluents containing heavy metals through the adsorption process. The low cost agricultural wastes such as sugarcane bagasse, rice husk, coconut husk, sawdust, neem bark, etc. have the potential for the elimination of heavy metals from wastewater.

8.15.1.5 Pyrolysis

In pyrolysis systems, agricultural waste is heated up to a temperature of 400-600° C in the absence of oxygen to vapourise a portion of the material, leaving a char behind. This is considered to be a higher technology procedure for the utilisation of agricultural wastes. Others are hydro-gasification, and hydrolysis. They are used for the preparation of chemicals from agricultural waste as well as for energy recovery. Agricultural wastes can be utilised to prepare alcohols for fuel, ammonia for fertilisers, glucose for food and feed by pyrolysis. Pyrolysis of agricultural waste yields oil, char and low heating value gas (Obi et al. 2016).

8.15.1.6 Direct combustion

Complete combustion of agricultural waste consists of the rapid chemical reaction of biomass and oxygen, the release of energy, and the simultaneous formation of the ultimate oxidation products of organic matter, CO_2 and water. The energy released is usually in the form of radiant and thermal energy provided oxidation occurs at sufficient rate; the amount of which is a function of the enthalpy of combustion of the biomass. If agricultural waste is to be utilised efficiently through thermal conversion process, there is need to fabricate these biomass wastes into solid form. It is usually burnt for heating, cooking, charcoal production, and the generation of steam, mechanical and electric power applications (Klass, 2004).

Selected References

Annadurai, K., Palaniappan, S. P., Chinnusamy, C. and Jayanthi, C. 1994. Integrated farming system: need of the hour. Kisan World 21(9): 39.

ASA. 1989. Decision reached on sustainable agriculture. Agronomy News January 1989. American Society of Agronomy, Madison, Wisconsin, USA. p. 15.

Baishya, A., Kalita, M. C., Mazumdar, D. K., Hazarika, J. P. and Ahmed, S. 2007. Characterisation of farming systems in the Borpeta and Kamrup districts of lower Brahmaputra valley zone of Assam. Journal of Farming Systems Research and Development 13(2): 168-175.

Balasubramaniyan, P. and Palaniappan, Sp. 2010. Principles and Practices of Agronomy. Agrobios (India), Jodhpur.

Bederska-Lojewska, D., Swiatkiewicz, S. and Muszynska, B. 2017. The use of Basidiomycota mushrooms in poultry nutrition - a review. Animal Feed Science and Technology 230: 59-69.

Behera, U. K. 2013. A Textbook of Farming Systems. Agrotech Publishing Academy, Udaipur. pp. 280.

Behera, U.K., Sharma, A. R. and Mahapatra, I. C. 2007. Crop diversification for efficient resource management in India: Problems, prospects and policy. Journal of Sustainable Agriculture 30(3): 97-217.

Bhan, S. and Behera, U.K. 2014. Conservation agriculture in India – problems, prospects and policy issues. International Soil and Water Conservation Research 2(4): 1-12.

Bhattacharyya, R., Ghosh, B. N., Mishra, P. K., Mandal, B., Rao, C. S., Sarkar, D., Das, K., Anil, K. S., Lalitha, M., Hati, K. M. and Franzluebbers, A. J. 2015. Soil degradation in India: Challenges and potential solutions. Sustainability 7(4): 3528-3570.

BIFAD. 1988. Environment and Natural Resources: Strategies for Sustainable Agriculture. Board for International Food and Agricultural Development Task Force, U.S. Agency for International Development, Washington DC.

CARDI. 2010. A Manual on Integrated Farming Systems. Caribbean Agricultural Research and Development Institute, Ministry of Economic Development, Belize.

CGIAR. 1978. Farming systems research at International Agricultural research Centres. TAC Secretariat, Rome, Italy.

CGIAR-TAC. 1988. Sustainable Agricultural Production: Implications for International Agricultural Research. Consultative Group for International Agricultural Research/ Technical Advisory Committee, TAC Secretariat, FAO, Rome.

Corsi, S. and Muminjanov, H. 2019. http://www.fao.org/3/i7154en/i7154en.pdf

Crosson, P. 1992. Sustainable Agriculture. Resources 106: 14-17.

Dale, V. H. and Beyeler, S. C. 2001. Challenges in the development and use of ecological indicators. Ecological Indicators 1: 3-10.

Das, A. 2013. Integrated Farming: An approach to boost up family farming. LEISA India 15(4): 1-5.

Davis, T. J. and Schirmer, I. A. (Eds.). 1987. Sustainability issues in agricultural development. Proceedings of the Seventh Agriculture Sector Symposium. The World Bank, Washington DC.

FAO and IDF. 2011. Guide to good dairy farming practice. Animal Production and Health Guidelines. No. 8. Food and Agriculture Organization of the United Nations and the International Dairy Federation, Rome.

FAO. 1977. Recycling of organic wastes in agriculture. FAO Soil Bulletin 40. Food and Agriculture Organisation, Rome.

FAO. 1988. Report of the FAO Council, 94th Session, 1988. Food and Agriculture Organisation of the United Nations, Rome.

FAO. 1996. Low External Input Sustainable Agriculture (LEISA) in Selected Countries of Asia: Regional Workshop Report and Case Studies. Food and Agriculture Organisation of the United Nations, Rome.

FAO. 2001. Conservation agriculture - case studies in Latin America and Africa. FAO Soils Bulletin No. 78. FAO, Rome.

FAO. 2001. Mixed Crop-Livestock Farming: A Review of Traditional Technologies based on Literature and Field Experience. Animal Production and Health Papers, 152. Food and Agriculture Organisation of the United Nations. Rome.

FAO. 2007. FAO-SARD Initiative. http://www.fao.org/sard/initiative

FAO. 2010. What is Conservation Agriculture? http://www.fao.org/ag/ca/1a.html

FAO. 2014. Building a common vision for sustainable food and agriculture: principles and approaches. Food and Agriculture Organisation of the United Nations Rome. pp. 18-31.

FAO. 2019. SDG Indicator 2.4.1: Proportion of agricultural area under productive and sustainable agriculture. Food and Agriculture Organisation, Rome. http://www.fao.org/sustainable-development-goals/indicators/241/en/

FAOSTAT. 2013. FAOSTAT database. Food and Agriculture Organization of the United Nations. Rome. http://faostat.fao.org/

Fard, S. H., Toghyani, M. and Tabeidian, S. A. 2014. Effect of oyster mushroom wastes on performance, immune responses and intestinal morphology of broiler chickens. International Journal of Recycling of Organic Waste in Agriculture 3(4): 141-146.

Francis, C. A. 1990. Sustainable agriculture: myths and realities. Journal of Sustainable Agriculture 1(1): 97.

Fresco, L. O. and Westphal E. 1988. A hierarchical classification of farm systems. Experimental Agriculture 24: 399-419.

Friend, A. and Rapport, D. 1979. Towards a comprehensive framework for environmental statistics: a stress-response approach. Statistics Canada, 1979.

Gangwar, B. and Singh. 2011. Efficient Alternative Cropping Systems. Project Directorate for Farming Systems Research, Modipuram, Meerut, India. pp. 339.

Gangwar, B., Ravisankar, N. and Prasad, K. 2012. Agronomic research on cropping systems in India. Indian Journal of Agronomy (Special Issue) 57: 171-181.

Ghosh, P. K., Bandyopadhyay, K. K., Wanjari, R. H., Manna, M. C., Misra, A. K., Mohanty, M. and Subba Rao, A. 2007. Legume effect for enhancing productivity and nutrient use-efficiency in major cropping systems - an Indian perspective: a review. Journal of Sustainable Agriculture 30(1): 59-86.

Ghosh, P. K., Nath, C. P., Hazra, K. K., Kumar, P., Das, A. and Mandal, K. G. 2020. Sustainability concern in Indian agriculture: needs science-led innovation and structural reforms. Indian Journal of Agronomy 65(2): 131-143.

Godi, N. Y., Zhengwuvi, L. B., Abdulkadir, S. and Kamtu, P. 2013. Effect of cow dung variety on biogas production. Journal of Mechanical Engineering Research 5(1): 1-4.

Gold, M V. 2007. https://www.nal.usda.gov/afsic/sustainable-agriculturedefinitions-and-terms-related-terms#term19

Hartung, J. 2003. Contribution of animal husbandry to climatic changes. In: Interactions between Climate and Animal Production. EAAP Technical Series No. 7. Wageningen Academic Publishers, Wageningen, The Netherlands.

Hayati, D. 2017. A Literature Review on Frameworks and Methods for Measuring and Monitoring Sustainable Agriculture. Technical Report No. 22. Global Strategy Technical Report. FAO, Rome.

Hazra, C. R. 2001. Crop diversification in India. In: Papademetriou, M. K. and Dent, F. J. (eds.). Crop diversification in the asia-pacific region. Food and Agriculture Organisation of the United Nations Regional Office for Asia and the Pacific, Bangkok, Thailand.

Hoang, Viet-Ngu. 2013. Analysis of productive performance of crop production systems: An integrated analytical framework. Agricultural Systems 116: 16-24.

Howlett, D. J. B. 1996. Sustainable Land Management in the South Pacific. Network Document No. 19. International Board for Soil Research and Management, Bangkok, Thailand.

http://14.139.252.116/agris/breed.aspx

https://agriinfo.in/classifications-of-cropping-system-645/

https://vikaspedia.in/agriculture/livestock/cattle-buffalo/breeds-of-cattle-buffalo

https://www.calsense.com/three-pillars-of-sustainability/

Hufnagel, J., Reckling, M. and Ewert, F. 2020. Diverse approaches to crop diversification in agricultural research - a review. Agronomy for Sustainable Development 40(14): 1-17.

Ibrahim, R. A. 2015. Tribological performance of polyester composites reinforced by agricultural wastes. Tribology International 90: 463-466.

Jayanthi, C., Mythili, S. and Chinnasamy, C. 2002. Integrated farming systems – A viable approach for sustainable productivity, profitability and resource recycling under low land farms. Journal of Ecobiology 14(2): 143-148.

Jayanthi, C., Rangasamy, A. and Chinnusamy, C. 2000. Water budgeting for components in lowland integrated farming systems. Agricultural Journal 87: 411-414.

Kassam, A., Friedrich, T. and Derpsch, R. 2010. Conservation Agriculture in the 21st Century: A Paradigm of Sustainable Agriculture. European Congress on Conservation Agriculture, 4-6 October 2010, Madrid, Spain.

Kassam, A., Friedrich, T. and Derpsch, R. 2019. Global spread of Conservation Agriculture. International Journal of Environmental Studies 76(1): 29-51.

Kassam, A., Friedrich, T., Shaxson, F. and Pretty, J. 2018. The spread of Conservation Agriculture: Justification, sustainability and uptake. International Journal of Agricultural Sustainability 7(4): 292-320.

Kessler, J. J. and Moolhuijzen, M. 1994. Low external input sustainable agriculture: expectations and realities. Wageningen Journal of Life Sciences 42(3): 181-194.

Klass, D. L. 2004. Biomass for renewable energy and fuels. In: Cleveland, C. J. (ed.) Encyclopedia of Energy, Vol. 1. Elsevier, San Diego. pp. 193-212.

Kumar, S. 2013. Integrated farming system models for food and nutritional security. In: Meena, M. S., Singh, K. M., Bhatt, B. P. and Kumar, U. (eds.). Gender perspective in integrated farming system, pp. 33-46. ICAR Research Complex for Eastern Region, Patna, Bihar.

Lal, R. 1991. Soil structure and sustainability. Journal of Sustainable Agriculture 1(4): 67-92.

Lal, R. 2013. Climate-resilient agriculture and soil Organic Carbon. Indian Journal of Agronomy 58(4): 440-450.

Lal, R. and Miller, F. P. 1990. Sustainable farming for tropics. In: Singh, R. P. (ed.) Sustainable agriculture: issues and prospective Vol. 1, pp. 69-89, Indian Society of Agronomy, IARI, New Delhi.

LEISA. 2006. Farming with biodiversity: the ecological basis for sustainable agriculture. LEISA magazine, Special issue 2006. http://www.leisa.info

Lenka, S., Lenka, N. K., Sejian, V. and Mohanty, M. 2015.Contribution of Agriculture Sector to Climate Change. In: Sejian V. et al. (eds.), Climate Change Impact on Livestock: Adaptation and Mitigation. Springer India, New Delhi.

Likens, G. E., Driscoll, C. T. and Buso, D. C. 1996. Long-term effects of acid rain: Response and recovery of a forest ecosystem. Science 272: 244-245.

Mackie, R. I., Stroot, P. G. and Varel, V. H. 1998. Biochemical identification and biological origin of key odour components in livestock waste. Journal of Animal Science 76: 1331-1342.

MacRae, R. 1990. A History of Sustainable Agriculture. https://www.eap.mcgill.ca/AASA_1.htm

Mahapatra, I. C. 1994. Farming system research - a key to sustainable agriculture. Fertilizer News 39(1): 13.25.

Mahapatra, I. C. 2010. Policy and methodological issues: Road map and strategy for integrated farming systems research in India. pp. 12-19. In: Gangwar, B., Varughese, K., John, J., Rani, B. Vijayan, M. and Mathew, T. (eds.). Manual on integrated farming systems, KAU Cropping Systems, Thiruvananthapuram & Project Directorate of Farming Systems Research, Modipuram, Meerut.

Mahfuz, S. and Piao, X. 2019. Use of medicinal mushrooms in layer ration. Animals 9(12): 1014.

Marshall, E. and Nair, G. 2009. Make Money by growing mushroom. Rural Infrastraucture and Agro-industries Division, Food and Agriculture Organisation of the United Nation. pp 64.

Meena, M. S. and Singh, K. M. 2012. Decision process innovations, constraints and strategies for adoption of conservation agriculture. http://dx.doi.org/10.2139/ssrn.2088710

Meena, M. S. and Singh, K. M. 2013. Information and communication technologies for sustainable natural resource management. http://dx.doi.org/10.2139/ssrn.2244751

Naresh Kumar, M., Murthy, C. S., Sesha Sai, M. V. R. and Roy, P. S. 2012. Spatiotemporal analysis of meteorological drought variability in the Indian region using standardised precipitation index. Meteorological Applications 19(2): 256-264.

Nguyen, T. A. H., Ngo, H. H., Guo, W. S., Zhang, J., Liang, S., Lee, D. J., Nguyen, P. D. and Bui, X. T. 2014. Modification of agricultural waste/by-products for enhanced phosphate removal and recovery: Potential and obstacles. Bioresource Technology 169: 750-762.

Obi, F. O., Ugwuishiwu, B. O. and Nwakaire, J. N. 2016. Agricultural waste concept, generation, utilisation and management. Nigerian Journal of Technology 35(4): 957-964.

Okigbo, B. N. 1991. Development of sustainable agricultural production systems in Africa. International Institute for Agricultural Research, Ibadan, Nigeria. 66 pp.

Okigbo, B. N. 1995. Major farming systems of the lowland savanna of SSA and the potential for improvement. In: Proceedings of the IITA/FAO workshop, Ibadan, Nigeria.

Panke, S. K., Kadam, R. P. and Nakhate, C. S. 2010. Integrated Farming System for suatainable rural livelihood security. In: 22nd national seminar on Role of Extension in Integrated Farming Systems for sustainable rural livelihood, 9-10th December, 2010, Maharashtra. pp. 33-35.

Patil, M. G., Rathod, P. K. and Patil, V. D. 2018. Compost: a tool for managing soil borne plant pathogens. International Journal of Current Microbiology and Applied Sciences 6: 272-280.

Pimentel, D., Hunter, M. S., LaGro, J. A., Efronymson , R. A., Landers, J. C., Mervis, F. T., McCarthy, C. A. and Boyd, A. E. 1989. Benefits and risks of genetic engineering in agriculture. BioScience 39: 606-614.

Pingali, P. L. 2012. Green Revolution: Impacts, limits, and the path ahead. Proceedings of the National Academy of Sciences of the United States of America 109(31): 12302-12308.

Plucknett, D. L., Dillon, J. L. and Vallaeys, G. J. 1987. Review of concepts of farming system research: What, Why, and How. In: Proceedings of Workshop on Farming Systems Research, February 17-21, 1986, ICRISAT, Hyderabad. pp. 2-9.

Pretty, J. 1996. Sustainable Agriculture: Impacts on Food Production and Challenges for Food Security. IIED Gatekeeper Series No. SA60.

Rangabhashiyam, S., Anu, N. and Selvaraju, N. 2013. Sequestration of dye from textile industry wastewater using agricultural waste products as adsorbents. Journal of Environmental Chemical Engineering 1: 629-641.

Reijntjes, C., Haverkort, B. and Waters-Bayer, A. 1992. Farming for the future: an introduction to low-external-input and sustainable agriculture. MacMillan Education Ltd., London. pp. 164-209.

RIRDC. 1997. Sustainability indicators for agriculture, Rural Industries Research and Development Corporation, Australia. 54 pp.

Sabiiti, E. N. 2011. Utilising agricultural waste to enhance food security and conserve the environment. African Journal of Agriculture, Food, Nutrition and Development 11(6): 1-9.

Sasikala, V., Tiwari, R. and Saravanan, M. 2015. A review on integrated farming systems. Journal of International Academic Research for Multidisciplinary 3(7): 319-328.

Senanayake, R. 1991. Sustainable agriculture: definitions and parameters of measurement. Journal of Sustainable Agriculture 1(4): 7-28.

Shaner, W. W., Phillip, P .E. and Schmehl W. R. 1982. Farming system research and development: guidelines for developing countries. West view Press, Boulder, and Colorado, USA.

Sharma, A. R., Jat, M. L., Saharawat, Y. S., Singh, V. P. and Singh, R. 2012. Conservation agriculture for improving productivity and resource-use efficiency: prospects and research needs in Indian context. Indian Journal of Agronomy 57 (IAC Special Issue): 131-140.

Signor, D., and Cerri, C. E. P. 2013. Nitrous oxide emissions in agricultural soils: a review. Pesquisa Agropecuária Tropical 43(3): 322-338.

Singh, C., Pathak, P., Chaudhary, N., Rathi, A., Dehariya, P. and Vyas, D. 2020. Mushrooms and mushroom composts in integrated farm management. Research Journal of Agricultural Sciences 11(6): 1436-1443.

Singh, J. P. 2010. Scenario of farming systems in different agroclimatic zones. pp. 20-41. In: Gangwar, B., Varughese, K., John, J., Rani, B. Vijayan, M. and Mathew, T. (eds.) Manual on integrated farming systems, KAU Cropping Systems, Thiruvananthapuram & Project Directorate of Farming Systems Research, Modipuram, Meerut.

Singh, R. P. and Ratan. 2009. Farming system approach for growth in Indian Agriculture. In: National seminar on enhancing efficiency of extension for sustainable agriculture and livestock production, December 29- 30, 2009, Indian Veterinary Research Institute, Izatnagar.

Sivakumar, M.V.K., Gommes, R. and Baier, W. 2000. Agrometeorology and sustainable agriculture. Agricultural and Forest Meteorology 103: 11-26.

Smaling, E. M. A. 1990. Two scenarios for the sub-Sahara: one leads to disaster. Ceres 126: 20-24.

Stern, W. R. 1993. Nitrogen fixation and transfer in intercrop systems. Field Crops Research 34(3-4): 335-356.

Subba Rao, A. and Mandal, K. G. 2007. Modern concepts of agriculture: Sustainable agriculture. Indian Institute of Soil Science, Bhopal. https://www.scribd.com/document/88818986/Revised-Sustainable-Agriculture

Swaminathan, M. S. 2010. Enhancing the disaster resilience of agriculture. The Hindu Survey of Indian Agriculture 2010: 7-10.

Swindale, L. D. 1988. Agricultural development and the environment: a point of view. Sustainable Agricultural Production: Implications for International Agricultural Research. TAC Secretariat, FAO, Rome. pp. 7-14.

Upadhyay, R. C., Singh, S. V. and Ashutosh. 2008. Impact of climate change on livestock. 36th Dairy Industry Conference, Varanasi, Uttar Pradesh, India.

Waheed, B., Khan, F. and Veitch, B. 2009. Linkage-Based Frameworks for Sustainability Assessment: Making a Case for Driving Force-Pressure-State-Exposure-Effect-Action (DPSEEA) Frameworks. Sustainability 1: 441-463.

Wanga, B., Donga, F., Chena, M., et al. 2016. Advances in recycling and utilisation of agricultural wastes in China: Based on environmental risk, crucial pathways, influencing factors, policy mechanism. Procedia Environmental Sciences 31: 12-17.

WCED. 1987. Our Common Future. World Commission on Environment and Development. Oxford University Press, Oxford.

Weil, R. R. 1990. Defining and Using the Concept of Sustainable Agriculture. Journal of Agronomic Education 19(2):126-130.

World Bank. 2020. Water in agriculture. https://www.worldbank.org/en/topic/water-in-agriculture

Subject Index